Electric Motor Manual

Other Electrical Construction Books of Interest

Electric Motor Manual:

Application, Installation, Maintenance, Troubleshooting

Edited by

Robert J. Lawrie

McGraw-Hill, Inc.

New York St. Louis San Francisco Auckland Bogotá
Caracas Lisbon London Madrid Mexico Milan
Montreal New Delhi Paris San Juan Singapore
Sydney Tokyo Toronto

Library of Congress Cataloging-in-Publication Data

Electric motor manual.

    Includes index.
    1. Electric motors — Handbooks, manuals, etc.
I. Lawrie, Robert J.
TK2511.E425   1987   621.46'2   87-4215
ISBN 0-07-036730-2

456789 HDHD 998765432

ISBN 0-07-036730-2

Printed and bound by Halliday Litho
The material herein is taken from articles originally published in
Electrical Construction & Maintenance magazine.

CONTENTS

Part III
TROUBLESHOOTING AND MAINTENANCE

Electric Motor Manual

INTRODUCTION

This volume is based on *EC&M's* popular Motor Facts department, which endeavors to present important and useful information concerning motors. The material has been prepared especially for facility (plant) electrical personnel, electrical installers, engineers, and others concerned with motor selection, installation, maintenance, and troubleshooting.

Ideas, techniques, and detailed methods are presented in this volume. Part I covers selection of motors from basic data to advanced criteria; Part II describes the how and why of proper motor installation; Part III discusses latest and most effective techniques of motor maintenance and troubleshooting.

Throughout the book, the most vital aspects are examined— energy-efficient motors, fundamentals of AC and DC motors, explosionproof motors, circuits, grounding, GF protection, why motors fail, testing and problem-solving techniques. Included are many photos, diagrams and sketches to aid fast understanding.

Robert J. Lawrie   *September, 1986*

PART I

SELECTION and APPLICATION

The AC induction motor—how it works

THE SQUIRREL-CAGE induction motor has long been the workhorse of industry because of its simplicity, rugged construction, and low manufacturing cost. With the increasing use of electronic adjustable-frequency controls, the AC induction motor appears to be well positioned to maintain its leadership. To obtain the best performance, it is important for the user to have a good understanding of motor and control operating principles and characteristics.

The induction motor is made up of two major components—the rotor and stator. The rotor consists of a structure of steel laminations mounted on a shaft, the squirrel-cage conductors or bars, end rings, and usually a fan. The stator has a formation of steel laminations mounted on a frame in such a manner that slots are formed on the inside diameter of the assembly. As shown in the accompanying photo, coils of wire or bars are made up and installed in the slots. The coils are connected to form a circuit such that when energized by an AC supply voltage, a rotating magnetic field is created in the stator.

This rotating magnetic field cuts through the rotor, inducing a voltage in the rotor bars, which in turn create their own magnetic field. The rotor magnetic field will attempt to line up with the stator magnetic lines of force. However, because the stator magnetic field is rotating, the rotor "chases" the stator field, never quite catching up, but following along just slightly behind. Essentially, this is motor action. Design of the motor will dictate the specific motor characteristics, such as horsepower, torque, speed, power factor, and duty cycle.

Motor characteristics

A 3-phase induction motor is furnished with three windings connected to a power source as shown in an accompanying sketch. When a polyphase alternating current flows in the stator winding, north and south poles are created in the stator, depending on how the windings are arranged and connected. The machine will always have at least two poles and a rotating magnetic field.

The speed (in rpm) at which the

induction motor rotates is dependent on the speed of the stator rotating field and is equal to $120 \times f/P$, where f is the frequency of the source (Hz) and P is the number of poles. For example, a motor having two poles supplied from a 60-Hz source will run at 3600 rpm; a 4-pole motor will run at 1800 rpm. The actual speed of the motor is slightly less because of "slip."

STATOR of AC squirrel-cage induction motor consists of steel laminations assembled to provide for slots that hold the stator windings as shown. Windings, or coils, are connected to create a rotating magnetic field when energized by an AC voltage.

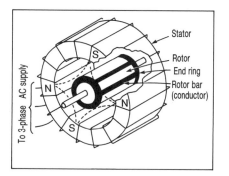

ROTATING MAGNETIC FIELD is created in stator by AC currents carried in stator windings. Three-phase voltage source results in creation of north and south poles that revolve or "move around" the stator. Magnetic forces in the rotor tend to follow the stator magnetic fields, producing rotary motor action.

Slip is the difference between the speed of the magnetic field and the speed of the rotor. Slip is necessary to permit the motor action to occur. Under load, the rotor slows down and the speed adjusts itself to the point where the forces exerted by the magnetic field on the rotor are sufficient to overcome the torque requirements of the load. The resulting speed is slightly less than the stator rotating field. For example, the actual speed of a 4-pole motor will be about 1725 rpm.

The slip necessary to carry full load depends on the motor characteristics. In general, the higher the inrush current, the lower the slip at which the motor can carry full load, and the higher the efficiency. The lower the inrush current, the higher will be the slip at which the motor can carry full load, and the lower will be the efficiency.

Motor supply voltages, current, torque, speed, and rotor impedance are closely related. By changing the resistance and reactance of the rotor, the characteristics of the motor can be changed; but for any one rotor design these characteristics are fixed.

An increase in line voltage decreases the slip, while a decrease in line voltage increases the slip. In either case, sufficient current is induced in the rotor to carry the load. A decrease in line voltage has the effect of increasing the heating of the motor; an increase in line voltage decreases the heating. In other words, the motor can carry a larger load. The slip at rated load may vary from 3% to 20% for different types of motors.

The locked-rotor current and the resulting torque are the factors that determine whether the motor can be thrown across the line or whether the current has to be reduced to get the required performance. Locked-rotor currents for different-type motors will vary from 2½ to 10 times their full-load current; but there are motors with even higher inrush currents.

Voltages higher than normal will increase the inrush current at a rate of about 12% for every 10% voltage increase, while the starting torque will increase at a rate of 20% for every 10% voltage increase. The reverse takes place if the voltage is lower than normal.

Torque and rotor resistance are closely related—the higher the rotor resistance, the higher the starting torque. This holds true only up to a fixed limit, beyond which a further increase in resistance causes the torque to decrease. The torque, however, is also affected by the flux in the air gap and the disposition and shape of the rotor slots and bars.

Speed-torque characteristics

Modifications in the design of the squirrel-cage motors permit a certain amount of control of the starting current and torque characteristics. These designs have been categorized by NEMA Standards into four main classifications.

1. Normal-torque, normal-starting-current motors (NEMA Design A)

2. Normal-torque, low-starting-current motors (NEMA Design B)

3. High-torque, low-starting-current, double-wound-rotor motors (NEMA Design C)

4. High-slip motors (NEMA Design D)

There are other variations, such as low-resistance motors with starting currents 8 to 10 or more times normal full-load current. These have a very high running efficiency and often find application where loads are driven continuously, such as on fans, milling machinery, pumps, and motor generators.

The basic NEMA types are derived from NEMA Standard MG1-1.16. The speed-torque curves for each type (NEMA Design) are shown in an accompanying drawing.

A Design A motor has a locked-rotor current that can be anywhere from 6 to 10 times full-load current. It has good running efficiency and power factor, high pullout torque, and low rated slip. The torque is about 150% at start. Pullout torque is over 200% of full-load torque.

A Design B motor has a higher reactance than the Design A motor, obtained by means of deep, narrow rotor bars. As a result, the starting current is held to about 5 times the full-load current. This motor allows full-voltage starting in some cases where the Design A squirrel-cage motor would require a reduced-voltage starter. The starting torque, slip and efficiency are nearly the same as for the Design A motor. Power factor and pullout torque are somewhat less, which may make this motor unsuitable

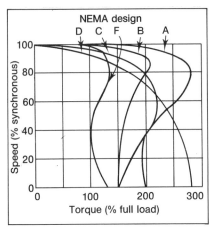

SPEED-TORQUE CURVES for NEMA-design motors

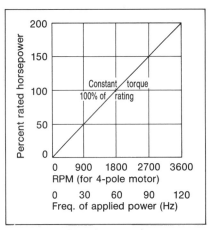

HORSEPOWER of motor increases as speed increases. However, torque remains constant, and it is essential that the ratio of voltage to frequency be kept constant.

for use where there are high load peaks. Design B is standard in 1- to 250-hp drip-proof motors and in totally enclosed, fan-cooled motors, up to approximately 100 hp.

The NEMA Design C motor has a higher starting torque than either the A or B design—about 200% of full-load torque. Breakdown torque, on the other hand, is lower than that for either the A or B design—about 180% of full-load torque.

Full-load torque is the same as that for both A and B designs. This type of motor has a "double-layer" or double squirrel-cage winding. It combines high starting torque with low starting current. Two windings are applied to the rotor, an outer winding having high resistance and low reactance and an inner winding having low resistance and high reactance. Operation is such

that the reactance of both windings decrease as rotor frequency decreases and speed increases. On starting, much larger induced currents flow in the outer winding than in the inner winding, because at low rotor speeds the inner-winding reactance is quite high. As the rotor speed increases, the reactance of the inner winding drops and, combined with the low inner-winding resistance, permits the major portion of the rotor current to appear in the inner winding.

The design has less current inrush than the standard type of squirrel-cage motor. The starting torque is rather high (200% and over), making this motor particularly adaptable to loads requiring high starting torque. However, the maximum pullout torque is lower than the starting torque, which may render this motor unsuitable where high load peaks are encountered. Typical applications are pulverizers, compressors, and conveyors.

The Design D motor produces a very high starting torque—approximately 275% of full-load torque. However, it has no true breakdown torque, as evidenced by the fact that torque decreases continuously with speed. It has low starting current, high slip, and low efficiency. Its slip is greater under load than other squirrel-cage motors and is used primarily where high starting torque is required but the running load is light or intermittent, such as for hoists and elevators. It is also frequently used for punch-presses and press brakes. The high slip of the motor makes the use of a flywheel very effective, since the motor speed changes with the load and allows the flywheel to store and give up energy during the operating cycle.

Induction motor current consists of reactive (magnetizing) and real (torque) components. The current component that produces torque (does useful work) is almost in phase with voltage. It results in a very high power factor that is close to 100%. The magnetizing current would be purely inductive, except that the winding has some small resistance, and it lags the voltage by nearly 90°. It has a very low power factor, close to zero.

The magnetic field is nearly constant from no load to full load and beyond, so the magnetizing portion of the total current is approximately the same for all loads. The torque current necessary to drive the load varies with the load, increasing as the load increases. At full load, the torque current is higher than the magnetizing current; and, for a typical motor, the power factor of the resulting current is between 85% and 90%. As the load is reduced, the torque current decreases, but the magnetizing current remains about the same—so the resulting current is lower in power factor. The less the load, the lower the load current and the lower the power factor.

Low power factor at low loading occurs because the magnetizing remains approximately the same at no load as at full load. However, a smaller magnetic field would be sufficient to provide the required torque at lower loads, and a very small magnetic field would be able to provide the small torque necessary to overcome friction at no load.

Changing frequency

An induction motor, as mentioned earlier, is a constant-speed device. Its speed depends on the number of poles provided in the stator. This assumes that the voltage and frequency of the supply to the motor remain constant.

Several methods can be used to vary the speed of an AC motor. The stator or primary winding can be connected to change the number of poles. For example, reconnecting a 4-pole winding so that it becomes a 2-pole winding will double the speed. This method can give specific alternate speeds but not gradu-

al speed changes. Or, the slip can be changed for a given load by varying the line voltage. However, torque is proportional to the square of the voltage, so reducing the line voltage rapidly reduces the available torque and will soon cause the motor to stall.

An excellent way to vary the speed of a squirrel-cage induction motor is to vary the frequency of the applied voltage. To maintain a constant torque, the ratio of voltage to frequency must be kept constant, so the voltage must be varied simultaneously with the frequency. Modern adjustable frequency controls perform this function. At constant torque, the horsepower output increases directly as the speed increases.

For a 60-Hz motor, increasing the supply frequency above 60 Hz will cause the motor to be loaded in excess of its rating, which must not be done except for brief periods. For a supply frequency of less than 60 Hz, the speed will be less than the design speed of the motor. As the frequency is reduced, the voltage should also be reduced, to maintain a constant torque. Sometimes it is desirable to have a high starting torque or to have a constant horsepower output over a given speed range. These and other modifications can be obtained by varying the ratio of voltage to frequency as required. Some control-

lers are designed to provide constant torque up to 60 Hz and constant hp above 60 Hz to permit higher speeds without overloading the motor.

The speed of an AC induction motor can be changed over a very wide range, from perhaps 10% to 20% of 60-Hz-rated speed up to several times rated speed. However, several cautions must be observed. At higher speeds, care must be taken not to exceed the hp rating of the motor. At speeds more than 10% above rating, the manufacturer must be consulted as to the ability of the motor to withstand the mechanical forces involved. At low speeds, roughly 20% of rated speed or less, especially if the motor is fan-cooled, care must be taken not to exceed the permitted motor temperature rise. If speed gets too low, the motor may "cog"—the rotor jumping from one position to the next instead of rotating smoothly—or it may stall completely.

Capability, versatility and flexibility of the AC induction motor is a matter of fact. To obtain maximum suitability and effectiveness when selecting and applying AC induction motors, many other factors must be considered. These include type of application, enclosure, mountings, coupling, bearings, insulation, temperature ratings, initial costs, operating cost, energy rating, and starting and control requirements.

DC motor fundamentals

DIRECT-CURRENT MOTORS are used extensively in industrial applications because of their ability to meet a wide range of torque and speed requirements. They are especially appropriate for applications requiring smooth acceleration over a broad range, accurate speed change and/or speed matching, and close control of torque or tensioning.

Although AC induction motors furnished with variable-frequency adjustable-speed drives are gaining in popularity, the DC motor will maintain its desirability for certain applications because of its special speed/torque characteristics. For example, AC motors driving heavy loads at about twice their rated torque will usually stall. Direct-current motors, on the other hand, can deliver about three times

their rated torque for short periods; and for very short periods, say 3 to 4 seconds, they can deliver up to five times rated torque.

The DC motor consists of two major components—the yoke or frame that contains the field windings and the rotating component called the armature. A very important part of the armature is the commutator and brush assembly.

The yoke, also called the stator, is a cylindrical frame of high-permeability iron alloy to which are bolted the pole structures. The poles are arranged in an alternating north and south pattern, and the frame provides a return flux path as well as mechanical support for the poles, bearings, and brush holders.

The armature consists of a shaft and

laminated punchings of silicon steel, slotted to accept the armature conductors or coils. The coil ends are connected to the segments of the commutator which in turn connects the armature conductors to the power source through the brushes.

Alternating current flows in the armature conductors, the frequency of which is a function of the speed of rotation and the number of field poles. The laminated armature of the poles reduces core loss. Direct current flows only in the circuits external to the machine, because the armature conductors are successively connected to and disconnected from the external circuits by the commutator.

Direct-current motors are classified in accordance with the types and connections of the various windings.

9

Shunt motors

The most widely used type is the shunt-wound motor. The name originated with early operation of these machines where the armature and field circuits were connected in parallel (shunt) to a constant-potential power supply. While the term "shunt" is still used, relatively few motors are now applied in this way. Shunt motors as now applied have their field circuits excited from a source of power separate from the armature power supply. The excitation voltage level is usually the same as the armature voltage, but special shunt-field voltage ratings of 15 to 600 V are available as a modification. The shunt motor is characterized by its relatively small speed change under changing load; rarely will the drop exceed 5%.

The speed of the shunt-wound motor can be changed by varying the shunt-field current or armature voltage. Speed control by changing the armature resistance is unsatisfactory because the speed regulation is objectionable. Whenever the load changes slowly, the flux changes as a result of armature reaction and speed will remain constant. However, if the load changes more rapidly than the self-induction of the field windings will allow the flux to change, then the speed will change rapidly. A speed range of approximately 4:1 with reasonable running stability is possible for loads up to full-load torque. Care must be taken never to open the field of a shunt-wound motor that is running unloaded. The loss of field flux causes motor speed to increase to dangerously high levels. Fig. 1 shows speed/torque characteristic of a DC shunt-wound motor.

Series motors

In series-wound motors, the field flux is created by coils that are electrically in series with the armature (see Fig. 2). When the motor starts, the current, and consequently the magnetic field, are at maximum values, producing a large starting torque. As the motor speeds up and the current is reduced, the field flux is reduced. The torque will vary as the square of the armature current, neglecting saturation of the field poles, which reduces this relationship. The torque and speed are very sensitive to the load current (which is also the field current) because of the corresponding change in flux. The speed of the series motor may be adjusted by shunting out the series winding, short-circuiting some field turns, or inserting resistance in series with the field and/or armature. However, speed adjustment is not easily accomplished. This motor has one disadvantage in that it tends to "run away" at light loads. The overspeed can reach a destructive value if the load is suddenly removed. For this reason, series-wound motors should be used only where the load is directly connected or geared to the shaft.

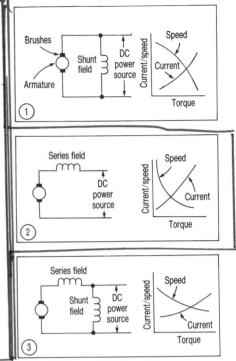

Compound motors

A motor built with both shunt and series fields is termed a "compound-wound" motor (Fig. 3). By proportioning the relative amounts of series and shunt windings, the designer may shift the motor characteristics to be more nearly shunt or more nearly series in nature.

Each winding has turns and wire sizes similar to the shunt-wound and series-wound motor field windings. The proportion of the total flux supplied by the series winding determines the amount of "compounding," which can be varied to suit the speed characteristics desired. A strong series field will give speed characteristics approaching those of a series motor. A weak series field will give characteristics approaching those of a shunt motor. Motors with series fields producing 40 to 75% to the total flux are often used, with a value of 50% most commonly provided. Compound motors having series fields producing 10 to 25% of the total flux are also used for some industrial applications. Generally, the speed characteristics lie between those of shunt-wound and series-wound motors. Compound-wound motors can be used when speed variation with load variation is permitted.

The starting torque of the compound-wound motor is high, although not so high as that of a series-wound motor. The torque will increase rapidly with load because the series field will increase the flux. The speed will decrease rapidly for the same reason. However, the motor will not run away at light loads because of the shunt-field flux.

The speed of the compound-wound motor can be adjusted with a shunt-field rheostat. Modern DC motor controls are somewhat complex; however they provide excellent control.

A motor has many torques

A BASIC UNDERSTANDING of torque and related terminology is essential to the proper selection and application of motors. This is important not only for new motors, but also when selecting energy-efficient motors for replacement purposes. Most authorities agree that the electrical man's knowledge of motor application must be broadened as newer designs replace existing motors. In the past, when a motor failed, it was repaired or a duplicate was ordered. Today, and in the future, the newer, energy-efficient motor will be selected.

Torque can be compared to "force." The term "force" is commonly used when describing linear motion. Torque describes rotary motion. Actually, torque is the product of force (lbs) and radius (ft) and is expressed in ft-lbs (or in.-oz). In other words, torque is a *twisting* force. Torque is a measure of the rotational force that a motor can produce.

A typical ac motor has many different kinds of torque. As noted on the accompanying motor-torque curve, the speed of the motor has a significant effect on its torque. The basic formula for horsepower (HP) shows the interrelationship of torque (T) and speed (rpm):

$$HP = \frac{T \times rpm}{5250}$$

(The value "5250" is a constant used

THOROUGH UNDERSTANDING of motor torque is essential to the proper selection of replacement motors, such as this 30-hp induction motor driving a large centrifugal screen. Belt-and-pulley drive is shielded by large steel base assembly. Also important is sizing of sheaves, which will have a significant effect on both speed and torque.

Motor speed/torque curve

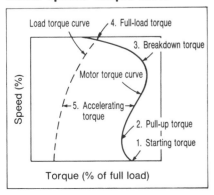

when torque is expressed in ft-lbs.) Starting torque, which is also called "locked-rotor" torque or "break-away" torque, is the torque developed at the moment the motor is started. (See point No. 1 in diagram.)

Pull-up torque is sometimes called minimum torque (point No. 2), because this is the speed range where minimum torque occurs. As shown, this area is often just above the starting-torque point on the curve.

Breakdown torque is the maximum value of torque exerted by a squirrel-cage induction motor without stalling. As shown by point 3, this is the maximum value reached by the torque curve. This characteristic is sometimes called "pullout torque," particularly when reference is made to synchronous motors.

Full-load torque is the torque that

the motor develops in producing rated horsepower at rated full-load speed (see point No. 4). In pounds, at a 1-ft radius, it equals the horsepower times 5250 divided by full-load speed.

Accelerating torque is the net difference at any speed between the torque required by the external load and that developed by the motor. This "extra" torque is shown by the shaded area between the motor torque curve and the load torque curve (point No. 5). It is this torque that accelerates the motor and the connected load. It is important that this torque be adequate to accelerate high-inertia loads to prevent the motor from stalling. As shown by the accelerating torque curve, the motor must develop more torque than the load requires at all times up to the final operating speed, where full-load torque is applied.

Selecting energy-efficient motors

DUE TO RAPIDLY rising power costs, there has been a growing demand for energy-efficient motors. Initially, the major emphasis has been the efficiency improvement at full load. But savings are also available at other operating conditions, since the motors have higher power factors, fewer PF-correction capacitors are required, and the branch-circuit losses are lower. These motors offer greater reliability, which will help to reduce operating costs and improve productivity. In addition, energy-efficient motors offer far greater application flexibility—standard energy-efficient motors can be used on many applications where special electrical design motors (normal efficiency) are currently being used.

Following is a summary of some of the many reasons why energy-efficient motors offer much more than a better way of converting electrical energy to mechanical energy or rotating motion.

Highest efficiencies. Energy-efficient motors have the highest efficiencies of any motor line designed or built. Some people ask, "Hasn't the industry really gone back to the good old U-frame design practices?" Actually, there is no comparison between the normal-efficiency U-frame, or T-frame motors and the new energy-saving motor efficiencies. A look at the history of 7½-hp TEFC 1800-rpm motor efficiencies, for example, shows that a 1944-design motor had an efficiency of 84.5%, a 1955 U-frame 87.0%, a 1965 normal-efficiency T-frame 84.0%, and a 1981 energy-saving T-frame 91.0%.

Lower operating cost. Price varies with horsepower, speed, enclosure and voltage. Operating cost will vary with electric power rates, operating hours, motor load and efficiency.

For a 75-hp 1800-rpm dripproof normal-efficiency motor operating continuously on only $0.05/kWh power, the operating cost will be over 13 times the first cost. The leverage is considerable. The price premium for an energy-efficient motor of this rating is approximately 22% and the power savings are approximately 4%. The payback in this case is 0.42 year or 5 months.

Lower demand charge. Frequently, utilities apply a demand charge for the maximum kilowatts used during the preceding 12-month period. Each motor operating during this peak 15- or 30-min period would contribute toward the charge. Only one 100-hp high-efficiency motor can reduce the demand by 2.9 kW. At a demand charge of $5.70/kW, this represents a monthly charge of $16.53 or $198.36 annually.

Fewer power-factor-correction capacitors. Energy-efficient motors have higher power factors as well as higher efficiencies. This means fewer kVARs of capacitors are required for power-factor correction. Savings are greater for smaller-horsepower motors.

Lower branch-circuit losses. Higher-efficiency motors have lower full-load currents. The energy saved in the motor branch circuits due to lower line currents is in addition to that realized within the motor. In a system with a maximum allowed voltage drop of 3%, the branch-circuit losses are 120 W less for a 50-hp energy-saving motor than a normal-efficiency motor. With power-factor-correction capacitors, the branch-circuit losses can be reduced another 180 W.

Lower no-load losses. Some motors are applied to varying loads or duty-cycle applications where they run for a period of time under a no-load condition. Even in these cases an energy-efficient motor provides savings. The percent reduction in no-load losses is significant with more-efficient motor designs.

Reduced air-conditioning load. As motors convert electrical energy to mechanical energy, they waste a certain amount of energy in motor losses or heat. If the motor operates in an area that must be air conditioned, such as a textile plant, these losses contribute an added load to the air-conditioning unit. Some users evaluate the savings of the more-efficient motor by adding an additional 25% for the savings in air-conditioning load.

Savings increase with time. The savings available by using more-efficient motors are directly proportional

to the power costs. Since power costs are forecasted to rise faster than most other costs, the value of energy-efficient motors will grow with each increase in the electric power rates.

Nameplate efficiency. There are many products being offered to customers as the ideal solution to their rapidly rising power costs. Although large potential savings are promised, the facts are vague. This is not the case with energy-efficient motors.

NEMA has adopted standard MG1-12.53b, which is an efficiency-labeling standard based on the probability-(bell-shaped)-curve concept that once the normal value of efficiency is established for a design, half the motors will be above and half below. The standard, which applies to NEMA designs A and B single-speed, polyphase squirrel-cage, integral-horsepower motors in the range of 1-125 hp, calls for the full-load nominal efficiency to be identified on the nameplate. This standard recognizes the variations in materials, manufacturing processes, test results, and motor-to-motor efficiency variations for a given design. The full-load efficiency for a large population of motors of a single design is not a unique efficiency, but rather a band of efficiencies. The new standard indicates the minimum and nominal efficiency to be expected from a motor design and a population of motors.

Any energy-efficient motor that complies with NEMA standards will have the nominal efficiency value on the nameplate. Some manufacturers also put their guaranteed minimum efficiency value on the nameplate.

Interchangeability. High-efficiency motors have the same hp-frame lineup as established by NEMA for normal-efficiency T-frame motors. Therefore, they have the same shaft height and mounting dimensions and are easily interchangeable. If they are used to replace a U-frame motor, it is necessary to use a transition base.

Conformation with NEMA standards. If a motor meets NEMA standards, the nominal efficiency will appear on the nameplate as indicated above. This efficiency will be determined in accordance with NEMA testing Standard MG1-12.53a, which is based on IEEE 112 Test Method B. In addition, these motors meet other NEMA standards with no compromise on inrush currents, starting or breakdown torques. Therefore, there are no specialized application rules or knowledge required in their application.

Protection and control. Since energy-saving motors use a conventional design approach and meet NEMA standards, they can be applied with conventional, NEMA-standard overload, short-circuit, single-phasing, reverse-phase, and stall protection. In addition, these motors employ the same control as a normal-efficiency motor when applied for reversing, plugging, overhauling loads, adjustable frequency, and other diverse applications.

Cooler and quieter operation. An added benefit of a motor with a higher efficiency is that there are fewer losses—less heat to dissipate. This results in a cooler-operating motor requiring less supplemental cooling normally supplied by the external fan on a TEFC design. Using a smaller or more-efficient fan can also reduce sound level.

Longer insulation life. Energy-saving motors operate below class B temperature rise, and most are below class A rise at rated load. These motors are built with either class B or class F insulation. The result is a reduction of operating temperatures of at least 20°C. Since the theoretical life of insulation doubles with every 10°C reduction in temperature, the insulation life increases up to four times the standard motor insulation life.

Improved bearing life. A common cause of bearing failure is too little lubricant. The longer the grease life, the less chance there is for bearing failure. Since the life of grease is directly affected by the motor operating temperature, the energy-efficient design offers greater reliability of the bearing system.

Lubricant life can be estimated from the bearing temperature. The temperature rise of bearings in a TEFC motor is approximately ½ to ⅔ of the winding

temperature rise. The energy-efficient design increases the lubricant life 200% compared to a normal-efficiency motor. Increased lubricant life will most certainly contribute toward greater bearing and motor reliability.

Less starting thermal stress. These motors also have less thermal stress during starting and short-term temperature peaks caused by abnormal operating conditions.

The rate of rise for an energy-saving motor is much lower than the rate for a standard motor. This results in lower total temperatures. The more-efficient design has a 30°C lower peak temperature after there are two overload trips from a cold start and a 51°C lower temperature after one trip following a hot restart. Since rapid changes in temperature coupled with long-term thermal aging are a frequent cause of motor failure, the more-efficient design is clearly a premium quality motor that will give greater reliability.

Greater stall capacity. The lower operating temperature and slow rate of rise of energy-efficient motors produce a greater stall capacity, providing better protection against failure due to single-phasing.

Less susceptible to impaired ventilation. The energy-efficient motor is less susceptible to thermal damage from loss of ventilation. Although the AC induction motor of normal-efficiency design is a very reliable device and requires little maintenance, it is essential that there be no obstructions of the cooling system. Since higher-efficiency designs operate at lower temperatures, they are less susceptible to damage from impaired ventilation.

Better buy than the old U-frame motors. When the motor industry shifted to the T frame with class B insulation, some motor users continued to purchase U-frame motors with their lower-temperature class A insulation. Although it is true that U-frame motors operated cooler, the T-frame normal-efficiency designs have equal or even greater reliability because the improved insulation system was designed to withstand higher temperatures.

Some motor users accepted class B insulation but insisted on class A rises

for additional insurance against failure available with this added thermal margin.

The energy-efficient T-frame motors offer cooler operation and extra thermal margin for significantly improved reliability. In addition, they have higher efficiencies, higher power factors, lower prices, lower noise, lighter weight, and longer life than the old U-frame motors. There simply is no comparison.

Higher service factors. To improve motor efficiency, the designer frequently adds additional electromagnetics. This increases the motor's overload capability in addition to its efficiency. The industry motor standard for service factor is 1.0 (zero overload) for totally enclosed fan-cooled motors and 1.15 (15% overload) for dripproof designs.

All dripproof and TEFC energy-saving motors have service factors of 1.15, and some even have thermal margins that would allow 30 to 40% overload. However, it is not recommended that energy-efficiency motors be operated continuously at these high service factors, because the efficiencies drop off at loads above nameplate rating and this could shorten bearing life and cause shaft breakage.

Motor users or specifiers who deliberately oversize motors to compensate for unknowns or cyclical loads can now more closely match the motor size to the load.

Better suited for energy-management systems. High-efficiency motors experience less stress from repeated starting associated with energy-management systems. Certain applications, particularly large fans and some compressors, have hard starting duty due to the considerable inertia that must be accelerated to full-load speed. Energy-efficient motors hold up much better than standard designs under these conditions.

Thermal margin for speed control. The added thermal margin is also helpful where motors are being used on inverter drives where the nonsinusoidal power supply produces extra losses and heat, and the motor is operating at low speed with less than normal cooling. The result can be a lower initial motor cost for the OEM.

Reducing motor losses

Energy can be conserved and power bills significantly reduced if motor losses are minimized. These articles will examine motor losses and cost-effective methods of reducing them.

Part 1—Motor losses and efficiency

MOTOR STATOR PRODUCTION LINE shows operators inserting coils into magnetic cores. Stator consists of many thin laminations stamped from high-quality magnetic steel, to minimize core losses. Slots are insulated, then coils are inserted into slots and connected.

REDUCING losses in electric motors presents a tremendous potential for saving electrical energy. Much effort, media attention, and federal and state regulations have been devoted to lighting levels and energy-efficient lighting, yet lighting accounts for only about 20% of the total electrical power consumption in the United States. Electric heating and miscellaneous applications account for about another 16%. By far the largest use of electric power is to drive motors—64%, or nearly two-thirds of the total electric power consumption, according to a 1976 report of the Federal Energy Administration (now the U.S. Dept. of Energy). Of this motor use, about 14% of the power consumed is for residential and other small motors, mostly single-phase, fractional-horsepower. The remaining 86% is for industrial, commercial and electric utility motors, almost all 3-phase, integral-horsepower (Fig. 1). Motors consume electric power equivalent to about 6 million barrels of oil per day—33% more than automobiles. Thus, any reduction in motor power consumption can be extremely valuable to the economy of the nation as well as saving the user substantial costs of motor operation.

Several approaches to minimizing motor losses may be used. Proper matching of the motor to the load—which does not always mean loading the motor to its full rating—can result in minimizing losses.* High-efficiency motors are now available having lower losses for a given load than standard motors, often saving enough in power to repay their cost premium in a very short time. A new invention by Frank Nola of the National Aeronautics and Space Administration (NASA) is now commercially available. Called a "power factor controller," it shows great promise of lowering motor losses. Another device, the "alternating-current synthesizer" (ACS), can also produce great economies where loads are not constant. The ACS provides variable frequency and voltage to a motor, adjusting the motor speed to match the required output. New technology is reducing the initial cost of the ACS variable-speed drive to a point where it is often a desirable choice.

We will examine the nature and measurement of motor losses and then evaluate the advantages, disadvantages, and economics of high-efficiency motors, the power-factor controller, and the ACS variable-speed drive as means for reducing these losses.

The nature of losses

The total power consumed by a motor consists of the power used for driving the load and power lost within the motor. The ideal motor would have power input exactly equal to the work done or power output, and be 100% efficient. Real motors waste some power. The losses for a typically loaded motor are about 10% of total power input for large integral-horsepower motors, 10%-15% for smaller integral-horsepower motors, and often more for fractional-horsepower motors. A lightly loaded motor will have higher losses as a percentage of input power than

14

Fig. 1. Electric power consumption in the U.S.

Fig. 2. Variation of efficiency with load for typical induction motor

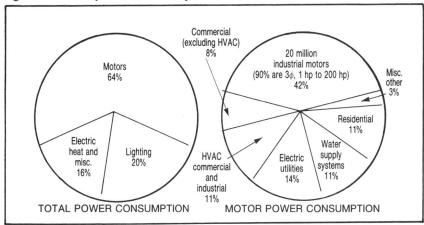

TOTAL POWER CONSUMPTION   MOTOR POWER CONSUMPTION

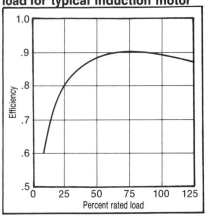

the same motor loaded to nearly its full rating, since the fixed losses are constant regardless of load. Since motors are rarely loaded to 100% of their rating, they are designed to be reasonably efficient when loaded from about 60% to 100%, with maximum efficiency in the 75%-80% range in most cases (Fig. 2). Efficiency is very low at light loads, and since an unloaded motor does no work but consumes some power, efficiency drops to zero at no load.

Power losses in a motor are that portion of the input power that becomes heat rather than driving the load. These losses can be divided into two basic categories—fixed losses and variable losses. *Fixed losses* are assumed to be constant at all conditions of motor loading from no load to full rated load. This is not exactly true, but it is nearly so, and little significant error is created by this approximation. Fixed losses include magnetic core losses (hysteresis and eddy current) and mechanical friction losses (bearing friction, brush friction, and air friction or windage). *Variable losses* are those that vary with the load on the motor and thus with the motor current. These losses increase as the load on the motor, and therefore the current drawn by the motor, increase. They are primarily the power lost in the resistance of the motor windings and are often called copper losses, or I^2R losses. Variable losses also include stray load losses, such as minor variations in fixed losses with load and speed and other small miscellaneous losses. Variable losses are approximately proportional to the square of the motor load current.

Motor efficiency is the output of the motor divided by the electrical input to the motor, usually expressed as a percentage. Power or work output is input less losses.

$$Efficiency\ (\%) = \frac{watts\ output}{watts\ input} \times 100$$

$$= \frac{746 \times HP}{E \times I \times PF} \times 100$$

$$= \frac{input - losses}{input} \times 100$$

Winding losses

Voltage applied to a motor drives current through the motor windings. In the typical 3-phase induction motor, voltage is applied directly to the stator, or primary winding, and induced current flows in the rotor, or secondary winding. The secondary winding of a squirrel-cage motor consists of bare conductors or bars, solidly connected together by a conducting ring at each end, producing a secondary winding that is part of the rotating armature. The squirrel-cage secondary has no external connections.

Wound-rotor induction motors have insulated coils in the secondary, with connections through slip rings to brushes and external terminals. A variable resistance or, more rarely, a separate voltage source, is connected to these secondary winding terminals to vary the slip and therefore the motor

speed. Some motors are designed to run at synchronous speed, and these have a wound secondary connected through slip rings and terminals to an external dc power source. The greatest number of motors, by far, are squirrel-cage induction type. In this discussion of losses and their elimination, we will consider mainly this type of motor.

Winding losses are often called "copper losses," but today this is not strictly correct. Many motors, especially fractional-horsepower and smaller integral-horsepower units, have stator windings of aluminum wire. Most motors today have armatures with the conductors, end rings, and fan blades cast of aluminum as a single unit. A more accurate term would be I^2R losses, indicating power converted to heat by the resistance of the copper or aluminum conductors. The total I^2R losses are the sum of the primary (or stator) I^2R losses and the secondary (or rotor) I^2R losses (including any brush and contact losses for wound-rotor motors).

The actual I^2R losses in the windings will depend not only on the current, but also on the winding resistance under operating conditions. The actual, effective resistance of a winding will vary with temperature, load, excitation, ac skin effect, magnetic influences, uneven sharing of current among conductors, and similar factors. Even for one specific set of conditions it is extremely difficult to determine the true values of winding resistance. Standard loss-measurement practice, therefore, uses as the resistance value the dc resistance at the steady winding temperature attained where the motor is run at full load in a 25°C ambient. Resistance cal-

15

culations are corrected to this temperature. Errors introduced by this assumption are compensated for, along with other small losses, by considering them as stray load losses.

Core losses

An induction motor operates by generating a rotating magnetic field, which causes the armature to rotate. To make the motor efficient and reasonable in size, the magnetic fields are concentrated and directed by high-quality magnetic steel in both the field (stator) and armature (rotor), with a minimal air gap between them. When the molecules in the steel are magnetized first in one direction and then in the other by the applied alternating current, there are energy losses in the steel. These show up as heat and are known as hysteresis losses. These losses increase with increasing magnetic flux density (higher current) and also with increasing frequency of the applied voltage.

The alternating magnetic field also induces small voltages in the steel core, which causes random currents to circulate within the steel. These are known as eddy currents, and acting on the resistance of the steel they also produce power loss that appears as heat. Eddy-current losses are minimized by making the steel cores of many thin laminations of steel, insulated from each other, in both the field and armature. Eddy-current losses vary with flux density and frequency of flux variation, as do the hysteresis losses, and these two are usually combined under the term *core losses*.

Core losses vary with the load current on the motor, speed changes, and other conditions of operation. It is very difficult to measure them under dynamic conditions. Therefore, the accepted method for determining core losses measures them under a standard set of conditions and assumes they are constant over the load range of the motor. Inaccuracies resulting from this assumption are accounted for under stray load losses.

Mechanical losses

Friction and windage losses are the mechanical energy consumed by the motor in overcoming bearing friction, the friction of brushes against commutator or slip rings (if any), and the friction of the moving parts, especially the cooling-fan blades, through the surrounding air. This lost energy is converted to heat, as are all other losses. Mechanical losses are assumed to be constant from no load to full load, which is a reasonable approximation but not absolutely accurate. As with core losses, any inaccuracies are accounted for under stray load losses.

There is no simple method of calculating mechanical losses. They must be determined by testing according to an accepted procedure. For those ac motors having brushes, testing is done with the brushes in place, on the slip rings, so that the total friction, including brush friction, is measured. For dc motors, brush friction against a commutator varies considerably with brush and commutator surface condition, so testing is done with the brushes lifted off the commutator. A standard wattage, depending on brush type and size, is added to the measured value to account for brush losses in dc motors.

Stray load losses

We have seen that several types of losses are assumed to be constant over the no-load to full-load range of the motor, although we know that they actually vary slightly with load. In addition, there are some losses that we cannot calculate. As a result of flux variation with load, skin effect, and conductor geometry, the current will be divided among the conductors somewhat unequally and will be less than perfectly uniform over the conductor cross-sections. Also, as current increases, conductor temperature will increase, causing an increase in conductor resistance and therefore in conductor losses. As magnetic flux increases with load current, there will be some increase in magnetic core losses. All these minor losses, from both known and unknown sources, are lumped together as stray load losses, which tend to increase with load. They

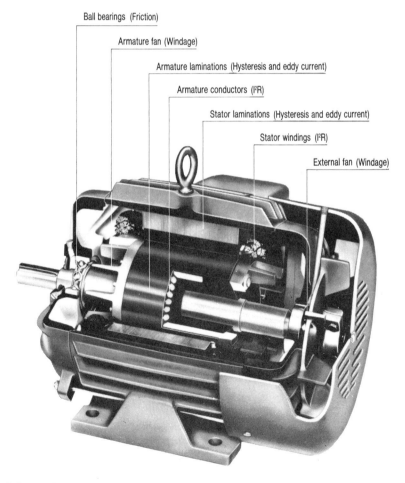

Ball bearings (Friction)

Armature fan (Windage)

Armature laminations (Hysteresis and eddy current)

Armature conductors (I^2R)

Stator laminations (Hysteresis and eddy current)

Stator windings (I^2R)

External fan (Windage)

INDUCTION MOTOR CUTAWAY VIEW shows important features and construction. Labeling indicates major components that contribute to motor losses and (in parentheses) the type of loss that takes place.

are determined by standard test procedures and calculations, and it is in the calculation of stray load losses that substantial variations in calculated efficiency can result.

Measuring efficiency and loss

It would seem that measuring efficiency should be a simple matter. Just load the motor, measure the mechanical power output, and measure the electrical input with a wattmeter. If we know input and output, efficiency can be calculated easily from the previously given formula. Unfortunately, things are not that simple and straightforward. A 1976 report to the Federal Energy Administration by the consulting firm of Arthur D. Little, Inc. stated,

Reliable and consistent data on motor efficiency is not now available to motor appliers. Data published by manufacturers appears to range from very conservative to cavalier.

In the absence of clear standards for efficiency testing, comparing data from different manufacturers was not a reliable means of choosing among motors. In fact, it penalized the conservative manufacturers and favored the overly optimistic manufacturers.

This situation has been improved considerably with the publication of IEEE Standard 112-1978 (revised from IEEE Standard 112A-1964) and with its adoption as an ANSI Standard by the American National Standards Institute. This document, entitled *IEEE Standard Test Procedure for Polyphase Induction Motors and Generators*, gives detailed procedures for measurement of motor efficiency (and other motor testing) as developed from theoretical and experimental information by the Rotating Machinery Committee of the IEEE Power Engineering Society. There are a number of other test standards in use throughout the world, but these generally are based on no-load tests and designated assumptions and calculations. Testing is faster and less costly than IEEE-112, but results tend to be overly optimistic estimates of efficiency (Table 1). The IEEE-112 Method B is difficult, and expensive, but it is the most accurate and thorough. Efficiencies are realistic, and savings will not be overstated. Note that on the same motors the three international standards gave apparent efficiencies from 2.0% to 4.7% higher than IEEE Method B. If efficiency is 80%, losses are 20%, and a 4% high efficiency measurement represents a 20% understatement of losses.

The IEEE standard lists five methods for determining efficiency. Method A uses an adjustable mechanical brake to load the motor to the desired torque. Method B loads the machine with a dynamometer to measure mechanical power output. Method C uses two duplicate motors, coupled together and connected to two sources of electrical power, one source having adjustable voltage and frequency to produce the desired loading. Methods E and F are known as "segregated-loss" methods and depend on measurement and calculation (Method E) or calculations based on the equivalent circuit of the motor (Method F). There is no Method D.

Method B is generally accepted to be the most accurate for motors in the range from 1 hp to 125 hp. With adequately rated test equipment, it may be used for even larger motors. It is based on dynamometer measurements with accuracy improved by segregated-loss determination and stray load loss measurement. Segregated-loss determination requires separate measurement and calculation of each type of loss—that is, friction and windage losses, core losses, stator I²R losses, rotor I²R losses, and stray load losses. Some of the greatest variation among motors can occur in the stray load losses, and the standard indicates ways to minimize errors and excessive optimism in determining these losses. The test procedures are clearly defined, and testing is done at full operating temperature or corrections are made for temperature differences. Required accuracy of instrumentation is detailed. Procedures are given to minimize or eliminate errors.

The test data and report form used

Table 1. Comparison of efficiencies

Standard	Full-load efficiency (%)	
	7.5 hp	20 hp
International (IEC 34-2)	82.3	89.4
British (BS-269)	82.3	89.4
Japanese (JEC-37)	85.0	90.4
US (IEEE-112 Method B)	80.3	86.9

EFFICIENCY TESTING of a 100-hp motor is a complex procedure. Motor is connected to a 300-hp dynamometer to vary and measure work output, and meters measure electrical input. Thermocouples, located inside motor and on case, are connected by many thin wires to strip-chart temperature recorder to measure temperature rise of windings, insulation and enclosure. Careful data recording permits accurate calculation of efficiencies.

Fig. 3 Test data for IEEE-112, Method B

Form B
Method B: Input-Output Test of Induction Machine

Type_____ Design _____ Frame_____ Hp_____ Phase_____
Frequency_____Volts_____ Synchronous r/min_____ Serial No_____
Degrees C Temperature Rise_____Time Rating_____ Model No_____

Test Point	1	2	3	4	5	6
(t_t) Stator-Winding Temperature, °C						
Ambient Temperature, °C						
Frequency, Hz						
Observed Slip, r/min						
*Corrected Slip, r/min						
Speed, r/min						
Torque,_____**						
(1) Dynamometer Correction_____**						
(2) Corrected Torque_____**						
(3) Power Output, hp						
Line Current, amperes						
Power Factor, percent						
Observed Power Input, watts						
(b) Stator I^2R Loss, watts, at (t_s) °C						
(c) Stator I^2R Loss, watts, at t_t						
(4) Input Correction = (b) − (c)						
(5) Corrected Power Input, watts						
(6) Efficiency, percent						

Performance Curve _____
*See 4.4.1. **Indicate torque units as N·m, or lb·ft.

Data Obtained from Performance Curve

Load, percent of rated	25	50	75	100	125	150
Power Factor, percent						
Efficiency, percent						
Speed, r/min						
Line Current, amperes						

t_t = temperature of stator winding as determined from stator resistance or by temperature detector during test
t_s = specified temperature for resistance correction (see 4.3.1)
(1) "Corrects" for windage and bearing loss torque of dynamometer and is equal to

$$\frac{(A - B)}{kn} - C$$

where:

A = power, in watts, required to drive machine when coupled to dynamometer with dynamometer armature circuit open
B = power, in watts, required to drive machine when running free and uncoupled.
C = torque output registered by dynamometer during test "A"
k = 0.1047 for torque in N·m
k = 0.1420 for torque in lb·ft
n = rotational speed in r/min.

(2) Corrected torque is equal to observed torque plus correction (1)
(5) This value is equal to observed power in watts plus correction (4).
(6) Percent efficiency = $\frac{(3)}{(5)}$ × 74 570.

for Method B is shown in Fig. 3. When test data is obtained in accordance with the methods called for in the standard, when the dynamometer is rated at not more than three times the motor horsepower (to minimize dynamometer error), when instrumentation is of the required accuracy, and when the proper temperature and other corrections are made, results should be accurate. In late 1977 and early 1978, to determine the reliability of efficiency testing as then being done by motor manufacturers, the National Electrical Manufacturers Association (NEMA) conducted a "round robin" test. Three motors, one each of 5 hp, 25 hp, and 100 hp, were sent to nine different motor manufacturers, who were requested to test for efficiency at various loads, using their version of Method B of IEEE-112. The results were shocking. Variations of several percent in efficiency were reported. This was equivalent to typical ranges of over ±15% on losses for the 5-hp and 25-hp motors, and ±20% to over 40% on the 100-hp motor, where total losses were smaller. The nine manufacturers were then requested to use the same test data, but to segregate losses, using "statistical smoothing" of the stray load losses. The results were distinctly improved, with the range of loss values reduced to about ±10% for the 5-hp motor, 12% for the 25-hp motor, and 15% for the

100-hp motor.

In late 1978, the three motors were again submitted to the nine manufacturers for retesting, with the input-output test method specified for use by all testers. The results were reported before and after segregated loss calculations. Even before segregating losses the results were noticeably better, with the range of losses about ±13% for the 5-hp and 25-hp motors, ±20% for the 100-hp motor. After segregating losses and smoothing stray load losses, the range was an acceptable ±6% on the 5-hp motor, ±8% on the 25-hp motor, and ±12% on the 100-hp motor.

This experiment resulted in the adoption by NEMA of a new standard, NEMA STD MG1-12.53, Paragraph (a) and (b), Revised January 17, 1980. This standard augments IEEE-112, specifying motor efficiency testing methods to improve the accuracy and reliability of results. This NEMA standard is now being accepted as an ANSI standard. It is anticipated that the next revision of IEEE-112 will also incorporate these new requirements for testing. Motor manufacturers who label their motors as NEMA designs are expected to test in accordance with this new standard, and it is highly recommended that for efficiency calculations, motor comparisons, and power-cost determinations that all data be based on IEEE-112, Method B, as modified by NEMA MG1-12.53 (a) and (b).

Power-factor effects

It must be pointed out that low power factor also can be a cause of high utility billing, if the utility has a power-factor penalty clause. Power-factor-correction capacitors can raise the power factor and eliminate or reduce these power-factor penalties.

Capacitors supply the reactive power consumed by the motor, which is *not* read by the kilowatt-hour meter and the demand meter. Therefore, although there may be some slight reduction in distribution-wiring I^2R losses as a result of the elimination of the reactive current for capacitors located at the motors (and not even this for capacitors located at the service equipment), capacitors *do not in any way reduce motor losses*. Although capacitors improve the power factor of the system feeding the motors, the power factor of the motor itself remains unchanged. The motor cannot tell whether its reactive power requirements are supplied by the power system or by capacitors.

Reducing motor losses

Energy can be conserved and electric power bills significantly reduced if motor operating losses are minimized. This article will examine losses and methods of reducing them.

Part 2—Methods, application, and payback

POWER losses in motors can be reduced by several different methods. High-efficiency motors, the NASA-developed power-factor controller, and variable-frequency, variable-speed drives are all effective when properly applied.

It is important to understand not only how each of these systems reduces losses and increases the efficiency of a motor, but also which method provides the most-effective results and the shortest payback on the investment.

High-efficiency motors

The most-obvious method of reducing motor losses is to make the motor itself more efficient. This can be done very successfully, but the more-efficient motor will also be more expensive. Twenty or more years ago, motors were considerably more efficient. In fact, for motors of 20-25 hp and below,

the standard motor efficiency was about equal to that of many of today's high-efficiency motors. Then better insulating materials were developed, making it possible to push more current through a given wire size and run it at a higher temperature. This, in turn, made it possible to reduce motor core sizes and make the motors smaller in overall dimensions. The net result was smaller, lighter, lower-cost motors, which became standard as a result of the competitive nature of the industry.

Until a few years ago, initial cost was the major factor in purchasing a motor of a given hp, design, and enclosure from among suppliers known to provide adequate reliability and availability. Operating costs, as determined by motor efficiency, were generally ignored. In the days of less than 2-cents-per-kWh electrical power, this was permissible, especially since in industry the cost of electrical power averaged less than 1% of the total value of products shipped. In addition, almost half of all motors were purchased by original equipment manufacturer (OEMs) to be incorporated into other equipment. The OEM is concerned with the reliability of his final product and its initial cost. He will purchase a motor of adequate reliability at the lowest cost and in most cases ignore the operating losses, which he does not pay and which the buyer of his equipment rarely considers.

This situation is changing. With power costs averaging 4 to 5 cents per kWh and still rising, the cost of electrical power can no longer be ignored. In addition, the availability of additional power is no longer assured as demand begins to exceed utility capacity, and it is essential to the national economy that waste be reduced. All these factors have made the high-efficiency motor desirable in spite of its higher initial cost. Today, the astute motor purchaser analyzes the overall cost of owning a motor, including operating costs, in selecting the motor, and even in purchasing equipment in which an OEM has installed motors. In many cases, the payback time for the increased cost

HIGHER MANUFACTURING COST of energy-efficient motor (left) compared to standard motor (right) is immediately apparent. The longer stator and rotor cores are obvious. Less visible are larger conductors in rotor and stator, optimized air gaps, new slot configurations, and other changes. Both motors are from the same manufacturer and rated 1½ hp.

SINGLE CORE LAMINATION of energy-efficient motor is checked carefully for concentricity by phototransistor gauge as part of program to minimize core losses. Not only must high-cost silicon sheet steel be used, but laminations must be of careful design, accurate in dimensions, and free of bends or nicks.

of energy-efficient motors can be short and, where applied correctly, very attractive economically.

Reducing winding losses

Winding losses are the I^2R heat losses resulting from current flowing through the windings. They represent 55% to 60% of the total losses. As we have seen, they vary with the square of the current in the stator windings and rotor conductors, and with the resistance of these windings. The current drawn by the motor is primarily a function of the load on the motor and cannot be reduced substantially, although improving the motor's power factor can reduce the current somewhat. Therefore, the best way to reduce I^2R winding losses is to reduce the resistance of the windings. This is what is done in energy-efficient motors, subject to limitations on physical size and cost.

In the stator, the size and number of conductors can be increased to lower the resistance. In recent years many standard motors in smaller sizes (up to 20 or 30 hp), which make up the numerical majority of motors purchased, have used aluminum wire in

their stator windings. However, increasing the cross section and number of the aluminum conductors would increase the physical size of the winding.

NEMA standard motor frame sizes have, for each frame, a standard dimension from the base to the centerline of the shaft. Increasing the size of the stator could require a larger dimension, making a larger NEMA frame size necessary. This is undesirable, since it increases both motor cost and space requirements. For these reasons, high-efficiency motors almost universally use copper conductors in the stator windings for minimum resistance in minimum dimensions.

Losses in the rotors of squirrel-cage induction motors can also be reduced by increasing conductor sizes. Most rotors of moderate-size integral-horsepower motors today have the secondary-conductor bars which make up the "squirrel cage" cast in one piece, along with the end rings and fan blades. In this application for ease of casting, aluminum is used. Rotors with copper bars, swaged and brazed into place, would be too costly to manufacture. By enlarging the slots, aluminum

bars of larger cross-section can be formed. Both slot size and geometric shape are very important in motor design. For wound-rotor motors, copper conductors can be used. Larger total rotor conductor cross-section decreases the I^2R rotor losses in either case.

A reduction in the air gap between stator and rotor and a reduction of magnetic flux density (by using more and better magnetic steel cores) will both reduce the magnetic field required, which will therefore reduce the current necessary to produce this magnetic field. It is the reactive component of the total motor current, the magnetizing current, that is reduced. This reactive component does not add to the power consumed; but it does increase the I^2R losses, which are read on the kWh meter. Reducing the reactive current thus cuts down the I^2R losses and also improves motor power factor.

In motors designed for minimum I^2R losses, each of these factors is used and balanced against the increased cost. A typical high-efficiency motor uses about 20% more copper and 15% more aluminum than the equivalent-hp standard-efficiency motor.

Reducing core losses

Core losses, as we have seen, consist of two components—hysteresis losses and eddy-current losses. They represent about 20% to 25% of the total losses. Hysteresis losses can be reduced by using silicon steel rather than carbon steel in the stator and rotor laminations. For a given magnetic flux density and steel thickness, the hysteresis losses are a function of the material used. Typical good-quality carbon steel will have losses of 4.5 to 5 watts per pound (W/lb) at a flux density of 15 kilogauss (kG). At the same flux density, even a low grade of silicon steel has losses of only 3.6 W/lb, and higher grades can have losses as low as 2.5 W/lb. However, silicon steels cost at least *twice* as much per pound as carbon steels—from 200% for the lowest grades to over 230% for higher grades—so the improvement in losses must be weighed against the increase in cost.

Eddy-current losses can be reduced by making the core laminations of thinner steel. For a medium-grade silicon steel, going from 24-gauge to 26-gauge laminations can reduce eddy-current losses by about 15%, and going

to 29-gauge laminations can reduce these losses by almost 20%. Material costs will increase by only about 5%, but there will be increased manufacturing costs as a result of the larger number of laminations required to make up cores of a given size.

Both hysteresis and eddy-current losses can be reduced by reducing the flux density. This can be accomplished by lowering the current in the windings, but only a small improvement is possible here, mainly by reducing the air gap between the stator and rotor. A greater improvement can be obtained by increasing the core size, since for a given total magnetic flux a larger core will result in a lower flux density as a result of the increased cross-sectional area. In addition, larger windings to reduce I^2R losses generally require a physically larger core. Increasing core diameter will increase the distance from the base of the motor to the center line of the shaft, requiring a larger NEMA frame size, so the increased core area is usually obtained by increasing core length, resulting in a longer motor. Greater motor length does not increase frame size. The decreased flux density reduces the

required magnetizing current in some measure, resulting in improved power factor and a small reduction in I^2R losses.

The reduction obtainable in core losses must be balanced against the increased cost of materials. Typically, a high-efficiency motor uses thin laminations of silicon steel, with about 35% more steel in the core than a standard motor of the same horsepower.

Reducing mechanical and stray-load losses

Mechanical losses are those resulting from friction and windage within the motor, representing only about 5% to 8% of the total losses. Friction is essentially bearing friction, and a slight improvement can be obtained by using high-quality, low-friction bearings. High-quality bearings are essential to maintain the closer clearances required by reduced air gaps between the stator and rotor in any case. Windage losses are caused by friction of the air against the rotating parts of the motor and of the cooling air against the internal cooling fans and circulating through the motor. Small improvements are obtained by optimizing fan-blade design and the paths of circulating cooling air. The possible improvement in efficiency is small, but the increased cost of better bearings and well-designed fan blades is also small. Energy-efficient motors use quality bearings internal fans, and air-cooling systems as a matter of course.

The term "stray-load" losses is a catch-all for some actual losses caused by leakage flux induced by motor currents, for variations in losses with load which, for convenience, are assumed to be constant, for losses resulting from nonuniform current distribution in stator and rotor conductors, and the like. They amount to about 11% to 14% of the total losses. Attention to the involved design of stator and rotor slot geometry and insulation is important. Little else can be done to reduce these losses per se, but motor improvements made to reduce winding and core losses and total motor current will also have the effect of reducing these losses. Careful manufacturing and handing of core laminations also will help. A well-designed high-efficiency motor will inherently have low stray-load losses.

Additional benefits

High-efficiency motors are selected when the savings in power consumption offset the higher initial cost. However, there are some other benefits that accrue to users. The lower-loss designs produce less heat and therefore run at lower operating temperatures. Some are manufactured with aluminum housings, which reduce operating temperatures as well as motor weight. These lower operating temperatures will result in longer motor life, since the life of insulating materials decreases as temperature increases. It is a rule of thumb that the life of insulation is doubled for every 10°C reduction in operating temperature. While insulation failure is not the only cause of motor failure, reduced insulation temperature will certainly increase average motor life, assuming that the insulating materials used are designed for the same or higher operating temperatures as those in standard motors.

Reduced operating temperatures will also improve the overload capacity of these motors. Since insulation is at a lower temperature to begin with, the motor can be overloaded for longer periods or greater amounts before the insulation reaches the maximum allowable total temperature. For the same reasons, these motors can be operated in higher ambient temperatures or at higher altitudes (where the thinner air provides less cooling) with no derating or less derating than standard motors. They will also tolerate wider variations in applied voltage without overheating. The lower operating temperatures should improve the effectiveness of lubrication and increase the life of motor bearings. Finally, the reduction in losses means less waste heat is produced. In an air-conditioned plant with large numbers of motors, this might reduce the overall cooking load. (Of course, in a plant that requires heating, the reduction in heat might increase the heat required from the heating system.) Generally speaking, high-efficiency motors can not only pay for their higher initial cost in power savings; they also should have longer life and a wider range of application than standard motors.

Table 2. Motor cost comparison

Basis of comparison	Standard motor	High-efficiency motor	Difference	Comments
Purchase price	$450	$562	$112	25% more
Efficiency	88%	92%	4%	4.5% higher
Losses	12%	8%	4%	⅓ less
Annual energy cost	$2713	$2595	$118	4.3% less
Annual cost of losses	$326	$208	$118	36.2% less
20-year energy cost	$54,260	$51,900	$2360	5¼ times motor cost
20-year cost of losses	$6520	$4160	$2360	21 times premium cost

Assumptions:
20-hp, 480-V induction motor
Cost of energy: 4 cents/kWh
Operation: 4000 hrs/year (2 shifts daily, 50 wks/yr)

$$\text{Annual energy cost} = \frac{0.746 \text{ kW/hp} \times 20 \text{ hp}}{\% \text{ efficiency} \times 0.01} \times \$0.04/\text{kWh} \times 4000 \text{ hrs}$$

Annual cost of losses = annual power cost × 0.01 × % losses

Economics and payback

As we have seen, producing energy-efficient motors with minimal losses requires more higher-priced materials, resulting in a more-costly motor—typically 20% to 25% more expensive than a standard motor. The reduction in losses will produce lower operating power consumption, and this must be sufficient to repay the increased initial cost within a reasonable period of time. At today's electric power costs, payback time for an energy-efficient motor is quite short, and it is difficult to understand the relatively slow penetration of these motors into the marketplace. For those applications with many hours of operation and high energy costs, only lack of knowledge can explain the purchase of standard motors over available energy-efficient motors. In 1980 sales of energy-efficient motors made up only 4 to 5% of the integral-horsepower motor market. The Department of Energy expects energy-efficient motors to make up about 33% of this market by 1985 and 74% by 1990.

Operating costs

It is important to recognize that motor operating costs are often many times the original cost of the motor. Let us contrast this with a typical family car of today, where purchasers are insisting on increased efficiency as measured in miles-per-gallon (as Detroit has learned at great cost). If a car is driven 15,000 miles a year and delivers 20 miles per gallon, it will consume 750 gallons of gasoline. At an average $1.25 per gallon, this amounts to $938 per year for fuel. If the car cost $7000, then the annual fuel cost is about 13% of the purchase price. By comparison, a 20-hp motor with an average standard efficiency of 88% costs about $450 and consumes about 17 kW of power. If this motor operates two shifts a day, 80 hours a week, for 50 weeks a year, it would run 4000 hours a year and consume 68,000 kWh. At an average 4¢ per kWh, the energy to operate this motor would cost $2720 a year—*over 6 times the purchase price* of the motor. Losses alone, at 12%, would cost $326 a year. In 1½ years the *cost of losses exceeds the price of a new motor.* (See Table 2.) An energy-efficient motor, with an efficiency of 92%, would have losses of only 8%, one-third less than the standard motor. This would result in a saving of $118 per year in energy costs, but the motor would cost $112 (about 25%) more. The energy-efficient motor would recover its additional cost in the first year of operation, which is not exceptional. Usually payback periods range from 1 to 2 years for the extra cost of energy-efficient motors. After that time, the power saving would continue for the life of the motor. In 20 years the saving would total 5¼ times the original cost of the motor and 21 times the premium paid for the energy-efficient motor.

Economic analysis

The preceding example was a very basic comparison of motor initial and operating costs to obtain an estimate of the time required by an energy-efficient motor to pay back its extra cost. Often this simple payback calculation is enough to justify the purchase of a more-efficient motor. In other cases, a more-comprehensive accounting approach may be desired, resulting in a true economic analysis for competing alternatives. This analysis will take into account the true cost of electricity, the time value of money, interest rates, income and investment tax considerations, and similar factors. Annual savings for the typical induction motor will vary with the duty cycle and power rate.

Several methods of evaluating the saving from energy-efficiency motors are shown in Fig. 4. The annual operating cost for each motor being compared is calculated first. The annual cost saving is the difference in operating cost between the lower-efficiency (or standard) and higher-efficiency motors. Even this simple calculation assumes a known and constant cost of electric energy. The true cost of electric energy, it must be understood, is not merely the charge per kilowatt-hour. It must also include demand charges, fuel-adjustment charges, power-factor penalties, and all other charges billed by the utility. If the power is onsite-generated rather than purchased from a utility, the true cost of that power is even more difficult to calculate. In determining costs over a period of years, allowance must be made for the increase in power costs with rising fuel prices and inflation. (If the utility charges a power-factor penalty, and the operating power factor is low enough to incur such a penalty, a wise investment would be PF correction capacitors, which would probably pay for themselves in less than two years, independent of any motor loss costs.)

Once the annual saving is computed, the simplest method of calculating the payback time is to see how many years (or months) it takes for these savings to equal or exceed the price premium paid for the energy-efficient motor. This method, while uncomplicated, ignores the fact that if this additional money were not spent on the higher-cost motor, it could be put to work earning interest or invested in production and earning profit. Therefore, especially at today's high interest rates and cost of money, the present value or present worth of these anticipated

Fig. 4. Evaluation of savings from energy-efficient motors

$$C = \frac{0.746 \times hp \times H \times P}{E}$$

$$S = C_{STD} - C_{EFF}$$

where

C = annual cost of operation (dollars)

$\quad C_{STD}$ = cost for standard motor

$\quad C_{EFF}$ = cost for energy-efficient motor

H = motor operating time (hrs/yr)

P = cost of electrtic energy ($/kWh)

E = motor efficiency (%/100)

$\quad E_{STD}$ = standard motor efficiency

$\quad E_{EFF}$ = energy-efficient motor efficiency

S = annual savings in operating cost using energy-efficient motors ($)

I. Simple payback

$$Y = \frac{D}{S}$$

where

Y = payback time (years)

D = difference in cost between energy-efficient and standard motors

II. Present worth of reduced operating cost ($)

$$W = S \times f$$

where

W = present worth of annual savings in operating cost ($)

$$f = \frac{(1 + i)^n - 1}{i(1 + i)^n} = \text{present worth factor}$$

(obtainable from compound-interest tables)

n = number of years of operation (the assumed motor life)

i = assumed annual interest rate (%/100)

III. Life-cycle cost of motor

$$\text{Life-cycle cost} = \text{initial cost} + W$$

(for assumed life of n years at interest i)

NOTE: Calculations are based on pre-tax figures and do not include effects of income taxes, depreciation, investment tax credits, and similar considerations.

Fig. 5. Cash flow and payback analysis

		ENCLOSURE	DP			
		CONSTRUCTION	AL			
		HORSEPOWER	125			
		SPEED	1800			
		VOLTAGE	460			

POWER COST   .0400 $/KWH
HRS OF OPERATION   8000.
TAX RATE   46.0%
DEPRECIATION   STRAIGHT LINE   – 10.0 YRS

A	B	C	D	E	F	G
0	-$316.17	–0–	–0–	–0–	-$351.30	-$351.30
1	$1070.98	$35.13	$1035.85	$476.49	$594.49	$243.19
2	$1156.66	$35.13	$1121.53	$515.90	$640.76	$883.95
3	$1249.19	$35.13	$1214.06	$558.47	$690.72	$1574.67
4	$1349.13	$35.13	$1314.00	$604.44	$744.69	$2319.36
5	$1457.06	$35.13	$1421.93	$654.09	$802.97	$3122.33
6	$1573.63	$35.13	$1538.50	$707.71	$865.92	$3988.25
7	$1699.52	$35.13	$1664.39	$765.62	$933.90	$4922.15
8	$1835.48	$35.13	$1800.35	$828.16	$1007.32	$5929.47
9	$1982.31	$35.13	$1947.18	$895.70	$1086.61	$7016.08

A—YEAR
B—SAVINGS
C—DEPRECIATION
D—EFFECT ON TAXABLE INCOME
E—EFFECT ON CASH FLOW FOR INCOME TAX
F—CASH FLOW AFTER INCOME TAX
G—ACCUMULATED CASH FLOW

PAYBACK PERIOD 0.59 YEARS, (ZERO INTEREST BREAK-EVEN POINT)
INTERNAL RATE OF RETURN 176.60% (DISCOUNTED RATE OF RETURN)

annual savings must be calculated. For example, if the saving equalled $100 per year for 10 years, the simple payback method would assume a saving of $1000. Taking interest into consideration and assuming a very low (for today) interest rate of 12%, the present worth of $100 per year for 10 years at 12% annually (from published compound interest tables) is only $565.

To get a realistic appraisal of the value of the power savings for an energy-efficient motor over a selected period of time, not only must the interest rate be predicted; the inflationary increase in the cost of power, the effect of taxes, investment tax credits, depreciation, and the like must also be factored into the calculation to some degree. This is difficult, but computer programs have been developed that do these complex calculations once the user selects his parameters. One manufacturer of energy-efficient motors has equipped his sales force with portable computer terminals. The salesman will come to you, connect his portable terminal to your telephone, and call his company's computer. You select interest rates, inflation rates, period of time, motor horsepower, annual hours of use and similar data, and the desired method of calculation, from the simplest to the most complex. The computer terminal will print out a complete annual savings, cash flow, and payback analysis. The computer program can

factor in depreciation methods, tax rates, investment-tax credits, fuel-cost increases and similar data. It can then provide such information as discounted-investment payback, life-cycle costs, and after-tax cash flow. A computer printout from this program is similar to that shown in Fig. 5.

Whether payback calculations are done simply or include all complex factors, it is apparent that where annual hours of operation are high, energy-efficient motors are a sound investment when a new motor is to be purchased. What is less obvious is that in many cases it is economically sound to replace an existing motor, especially one which, while satisfactorily operational, has been fully depreciated. In addition to the substantial saving in electric energy gained by replacing the motor with a more-efficient one, the new motor can be depreciated over its new life, and some investment tax credits may be available. The computer program can make it possible to examine the economics of a situation such as this.

When a motor fails, it can either be replaced or, as frequently is the case for larger motors, rewound. Many industrial users today have decided that all replacement motors purchased shall be of the energy-efficient design. Again, the computer program can help in deciding whether to rewind the motor or replace it with a more-efficient unit. One important consideration in this decision is that rewinding a motor may cause a lowering of its efficiency, increasing the losses. It has been found that in heating the core to remove the old coils sometimes excess heat is used. This can deteriorate the insulating surfaces between the laminations and substantially increase eddy-current core losses.

A major rubber company had many identical 30-hp motors in one plant, and when one failed it was replaced by a spare and then rewound. It was noticed that these originally identical motors were drawing different currents under the same load, and study showed that the current increased with each rewinding as efficiency decreased. This company decided that the increase in losses made it no longer economical to rewind these motors and replaced them with new motors as they failed. Remember that an increase of even 1% or 2% in losses can cost a tremendous amount over the life of the motor—often many times the cost of the motor itself. This small increase in losses may not be noticed as increased current in the rewound motor. The Electrical Apparatus Service Association (EASA) is doing some testing to determine the severity of this problem, since little data is available. Use of minimum heat and maximum care in removing old coils, protecting the core from damage, will result in the least effect on efficiency of the rewound motor but will increase the cost of rewinding. Motor rewinding shops usually are not equipped to measure the efficiency of a motor that has been rewound.

Purchasing an efficient motor.

In deciding whether to purchase an energy-efficient motor, it is important to make valid comparisons. In actually purchasing the motor, it is important that the desired efficiency level be achieved. This is still not simple, although recent standardization of measurement methods has improved the situation. It is extremely important, as explained in Part 1 of this series, that efficiency be measured by IEEE Standard 112, Method B, as modified by NEMA Standard MG1-12.53(a) and (b) in all cases.

Defining efficiency

The word "efficiency" applied to motors must be very carefully defined, as we saw in Part 1. Let us look at some of the terminology in use.

Nominal or *average expected* efficiency is the average value of a large number of motors of the same make and model. Individual motors can vary widely from this average, and this efficiency cannot be relied upon in economic determinations.

Calculated efficiency is just that— calculated from motor parameters and numerous assumptions. It often bears little resemblance to measured efficiency and should not be used for loss determinations.

Apparent efficiency is the product of motor power factor and efficiency. A guaranteed apparent efficiency tells the motor user nothing reliable in figuring costs, since it could come from a high power factor and low efficiency or from a low power factor and high efficiency or from something in between.

Minimum or *expected minimum* efficiency is a little more sharply defined. All motors should be equal to or higher than the efficiency value

Fig. 6. NEMA standard nameplate marking values

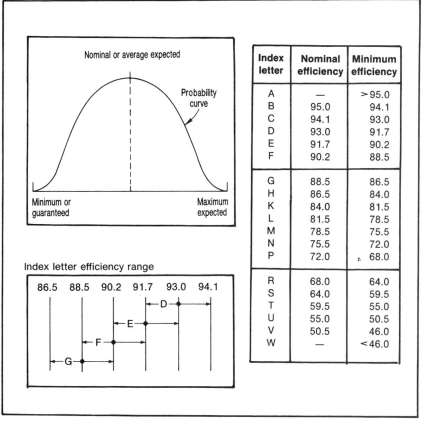

EXAMPLES of ranges of index letters D, E, F and G show nominal efficiency (dot), minimum permitted efficiency (left arrow), and maximum expected efficiency (right arrow).

given. However, there can be some question even here. One motor manufacturer defines "minimum" as the value that 95% of his motors will meet.

Guaranteed minimum efficiency is the value that the manufacturer guarantees all motors of that rating will meet or exceed. At least one manufacturer warrants that if his motor does not meet the guaranteed efficiency, he will replace or repair the motor or pay the user a cash sum to compensate for the additional losses.

When evaluating different motors, these guaranteed minimum values must be compared to obtain reliable cost figures. Care must be taken not to compare minimum values with nominal, average, or nonguaranteed values. One or two percentage points of error in efficiency can result in thousands of dollars difference in savings over the life of a motor.

NEMA standards

Normal variations in materials, manufacturing processes, and testing

of motors result in a range of efficiencies for a large batch of motors of the same design. Recognizing this, NEMA Standard MG1-112.53 (b) has been established. It is based on the normal statistical bell-shaped probability curve, which assumes that for a given design half the motors will be above and half below the nominal or average efficiency. The manufacturer must measure the efficiency by means of IEEE 112 Method B, as modified by NEMA MG1-12.53(a) and (b). He may then put a code index letter on the motor nameplate (Fig. 6), which will indicate both the nominal efficiency for that motor design and the minimum efficiency for any individual motor of that design. Some manufacturers mark the nameplate with the efficiency in percent, rather than with a code letter. In this case it is extremely important that the user determine whether this value is nominal or minimum efficiency. A manufacturer should, in any case, be willing to guarantee the minimum efficiency of his motor.

Effect of motor efficiency on energy savings

By DONALD ALDWORTH, Energy-Efficient Motor Specialist, Gould Inc., Electric Motor Div.

ENERGY-EFFICIENT MOTORS are designed to save operating expense over their useful operating life. Because these motors generally demand a premium price over alternatives, it is necessary to justify the investment by conducting a comparative analysis. Typically, an evaluation is developed from readily accessible data, because it is not economically feasible to perform an in-depth analysis every time a purchase decision is made. The evaluation is based on data supplied by motor manufacturers and assumptions pertaining to the operating conditions of installation.

Assuming the evaluation indicates that a high-efficiency motor is indeed the best buy, how can it be determined if the decision is correct? The proof can be obtained only by on-site testing. Is the motor performing as expected, and is it saving energy in accordance with calculations?

A review of selected installations with favorable results lowers the risk on future purchasing decisions. On the other hand, if follow-up analysis results are unfavorable, either original assumptions were wrong or the high-efficiency motor is not performing as expected. More than 100 in-plant watt-hour-meter tests, under a variety of applications, indicate that a favorable trend has developed.

What's the best buy?

Best-buy evaluation decisions are typically developed from the most readily available data. In-depth analysis, taking into account all potential variables, simply requires too much time.

The efficiency and power-factor performance of an energy-efficient motor are obtained from vendor literature. The *nominal efficiency* value represents performance that is the average of a large population of motors of the same design.

For comparative analysis, the next step is to determine the efficiency of the *other* motor. In most cases, the other motor is a standard-efficiency-design motor, perhaps one that is already installed and operating. Determining the actual efficiency for this motor is a difficult task. Therefore, the performance data are often obtained from "industry average" tables. Short of submitting the standard motor to a qualified test facility, using the average table is sufficient for most initial evaluations.

Armed with performance data and prices of standard and high-efficiency motors, an evaluation of the best buy requires estimation of operating time and energy cost. These variables are functions of the application and input obtained from the user.

The most basic evaluation requires the following minimum information:

Efficiency—a measure of how well the motor converts electrical energy input into mechanical work output (watts in ÷ watts out).

Operating time—the number of hours per day, week or year the motor is expected to operate.

Energy cost—the total cost of energy ($/kWh) including the basic rate charge, power-factor penalties, demand, and fuel-adjustment charges. It is obtained by dividing the total kilowatt hours into the total utility charge.

EXAMPLE

Assume that two 40-hp motors—one a standard motor (A) with a stated efficiency of 89.5%, the other (B) with a 92.4% efficiency—are to be operated 6000 hrs/yr. The total cost of energy is $0.05/kWh. Savings realized by the increased efficiency of motor B are given by the equation

$$\text{Hp} \times .746 \times \text{N} \times \$/\text{kWh} \times \left(\frac{1}{E_B} - \frac{1}{E_A} \right)$$

Substituting the given data in the equation results in a savings of $314/yr.

This evaluation is only as accurate as the input data. The efficiency input data are derived from statistical averages. No two motors are created equal. Manufacturing variances give each motor a personality of its own. In practice, similarly rated motors will actually have efficiencies within a tolerance band centered around the average or nominal. Through various sampling tests, NEMA has determined that a deviation may be expected of ±20% of motor losses and subsequently developed a standard that provides a common means of testing and labeling motor efficiency. NEMA MG 1-12.53 a and b establishes a common language for comparing motors made by various manufacturers. It defines the method of testing, interpretation of test results, and nameplate marking. Testing and labeling standards apply to new-motor production.

Factors affecting motor performance

From a simplistic point of view, a motor is used to convert electrical energy into mechanical rotating energy (power). In the laboratory, motor

Fig. 1. Saturation curve

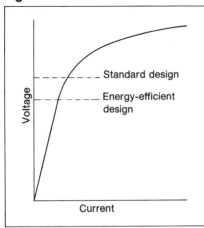

designs are evaluated for their effectiveness by the monitoring of several key parameters under ideal power-supply conditions. The results are compiled in published literature and marked on nameplates to be used in the selection of motors for actual job applications.

When installed on a jobsite, the motor is subjected to an environment different from the laboratory thus motor performance on the jobsite is different from its performance under laboratory conditions.

The power supply to which an alternating current motor is connected directly affects the motor's operating performance. Variations in voltage, frequency and load will affect motor efficiency and power factor.

Effects of voltage variation. NEMA Standard MG1-12.43 states that alternating-current motors shall operate successfully under running conditions at rated load with a variation in voltage (at the motor terminal leads) up to plus or minus 10% of the design value. In the interest of a more flexible design for general application, NEMA Standards define the voltage variation that all motors—standard or high efficiency—must meet.

A differentiation must be made between system voltage and applied voltage. *System voltage* is supplied to users by an electric utility or by in-plant substations through transformers and can usually be adjusted by taps. (For example, plant system voltage = 480/240 V.) *Applied voltage* (motor nameplate voltage) is the voltage at the motor terminal leads, which allows for the I^2R or voltage drop through distribution lines in plants or buildings. (For example, the industry standard at motor terminals = 460/230 V.)

Much of the effect of voltage variation at the motor terminals depends on how hard the motor design is working the magnetic iron. If the motor's design is working "light" magnetically, the volt-versus-amp relation is basically linear. Most energy-efficient motors are designed this way. On the other hand, if the motor's design runs close to saturation, the relation is not linear. Current will increase in greater proportion to an increase in voltage. (See Fig. 1.) Many standard T-frame motors are designed to operate at full load, near the saturation point.

The efficiency of a motor operating at full load is not likely to change much with small voltage variations, because stator and rotor I^2R losses tend to move in a direction opposite to the core loss. Thus, there is little cumulative change. At other than full load, however, variations in voltage result in a more significant loss of efficiency, as reflected in Fig. 2. Energy-efficient motors, due to their design and construction, are generally more tolerant to voltage variation. Notice the difference between the motor types, the standard motor being influenced in the normal load range and the energy-efficient motor being affected only at loads above 100%.

Since magnetizing current increases with voltage, the power factor of an induction motor becomes lower at voltages higher-than-rated. At lower voltages, the power factors will generally show improvements.

As with efficiency, power factor will change more drastically with voltage variations with standard motors than it will with energy-efficient motors.

Reduction of motor efficiency due to high or low voltage applied to the motor terminals can be corrected in the user plant by a number of established engineering methods. The simplest action would be to adjust the tap settings on the appropriate transformer. Most low voltage problems are caused by excessive voltage drop in the lines feeding the motor. If the loads vary considerably over the course of a typical day, automatic tap-changing equipment may be required. If the motor is at the end of a long feeder run, voltage drop may be reduced by rearranging the system. Since the cause of voltage drop is current (I^2R), the reactive component of current draw can be reduced by correcting power factor.

Voltage unbalance. Up to now, we have assumed the application of a *balanced* 3-phase voltage above or below the nameplate value. Another variation of particular importance to the smooth and efficient operation of an induction motor is voltage unbalance. The % phase-to-phase voltage unbalance is best illustrated by the relationship

$$100 \times \frac{\text{deviation from average voltage}}{\text{average voltage}}$$

When the line voltages applied to a polyphase induction motor are not exactly the same, unbalanced currents flow in the stator windings. The effect of unbalanced voltages on motors is a "negative sequence" voltage having a rotation opposite to that of balanced voltages. Unbalanced voltage produces a corresponding negative-sequence flux

Fig. 2. Effect of line voltage on efficiency

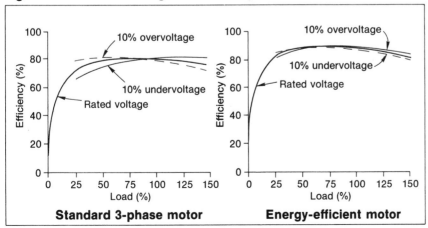

causing unbalanced current in excess of those under balanced voltage conditions. The level of full-load current unbalance will generally be on the order of six to ten times the voltage unbalance, and this increases as the load decreases. (See Fig. 3.)

The efficiency of the motor falls off dramatically as the voltage unbalance increases. For example, a 3.5% voltage unbalance will result in about a 25% increase in losses and corresponding temperature rise. Also, even a relatively small unbalance will cause the temperature of the motor to rise rapidly. In the phase with the highest current, the percentage increase in temperature rise will approximate two times the square of the percentage voltage unbalance. Such an unbalance can also result in electromechanical vibration, leading to bearing failures.

One common cause of unbalanced voltage conditions is an open circuit in the primary of the distribution system. Unevenly distributed single-phase loads on the same power-supply system is another cause of phase-to-phase unbalance.

Whatever the cause, it should be corrected. Excessive current will result in increased motor losses, deterioration of efficiency, and shortened motor life.

Load. Most electric motors do not operate at rated horsepower, which means that the nameplate efficiency, power factor and current do not apply. Nameplate values are generally assumed to be at "full load" or "rated" conditions. Data sheets covering various load points are usually available from manufacturers. Fig. 4 illustrates the effect on efficiency and power factor as the load changes for a typical motor.

Efficiency is essentially constant over a range from 50 to 125% of full load. Power factor is more severely affected by underloading. By comparison, energy-efficient motors have similarly shaped curves but their efficiencies are generally higher for any load point. Of course, every motor design has its own unique load curves, with peak efficiency occurring at less than, or greater than, full load.

Thus, when evaluating potential energy savings of motors with different designs, it is best to perform the evaluation with efficiency and power factor values corresponding to the anticipated loading of the specific motors. For example, assume a pump application requires a 40-hp, 4-pole motor. The

Fig. 3. Effect of voltage unbalance on current

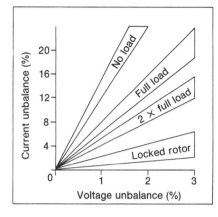

Fig. 4. Typical standard motor characteristics

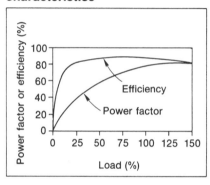

Fig. 5. Example of duty cycle

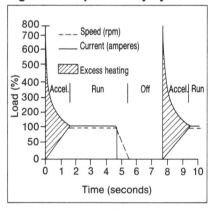

efficiency of a standard motor is 89.3% and an energy-efficient design is 92.0% at full load. At continuous operation, the savings will be 8590 kWh per year. In reality, however, the motor is operated at ¾ load, and the respective efficiencies at this load point are 88.0% and 92.4%. The calculated annual savings jumps to 14,145 kWh per year.

Duty cycle is another consideration of load that infers varied loads over time. Usually the duty cycle is repetitive and can be represented by a time graph. Strip recorders displaying current or watts versus time are examples of instruments used to obtain duty-cycle information.

If a duty cycle requires the motor to be shut off for part of its cycle as shown in Fig. 5, then excessive temperature rise may become a problem. Starting current will generate heat energy (losses) in proportion to the square of the current. On the other hand, a motor's ability to dissipate heat by ventilation is proportional to the square of the speed.

Motors applied to duty-cycle loading do not experience constant load, and therefore the efficiency performance is also varying over time. The average load and therefore average efficiency can be approximated; but remember that an idling motor performs no useful work but may be drawing 30% of its full-load kVA.

Other variables. Other environmental conditions affect efficiency. Although harder to measure, they can have significant influence over motor performance. In general, these influences fall into the category of maintenance. Poor maintenance will ultimately shorten motor life. The most-common result is premature failure because of excessive temperature rise. Bearings and insulation life are most vulnerable to excessive heat. Heat is energy and therefore exactly the sort of motor loss to avoid.

There are a number of common abnormalities that have an adverse effect on a motor's performance. These include mechanical misalignment, improper bearing lubrication, improper V-belt application, and ambient temperature and ventilation.

Another area of concern is the effect of rebuilding/rewinding a failed motor. The process of burning out the old windings has the potential of increasing core losses. If the burnout temperature is excessively high, above 750°F, the magnetic properties and interlaminate insulation may be degraded and result in a motor with lower efficiency after rewind.

Case studies in motor efficiency

Voltage variation, voltage unbalance, and loading affect the efficiency performance of all motors. These factors represent the primary external influence reacting with the internal design. Common designs will react in similar ways. Motors of different electrical designs, such as standard and high-efficiency motors, will react in a dissimilar manner to the same external stimuli.

From numerous studies, it appears that variations in external conditions will generally have a more detrimental effect on standard-efficiency designs than high-efficiency designs. The three case histories reported here are typical examples that support this trend.

The case histories are the results of watthour-meter tests conducted on the job under operating conditions. A cumulative watthour meter with elapsed-time meter was wired into the motor circuit being tested. Test data were collected by the companies involved. The cost of energy used in these examples represents a nationwide average. The purpose of reviewing these case studies is to demonstrate the performance difference between calculated (or expected) results and actual test results. In each case, unless otherwise noted, the energy-efficient motor (Motor B) substituted for the standard motor (Motor A) had the same ratings.

CASE 1. Chemical processing plant

Application	Screw compressor
Motor type	100 hp, 1800 rpm, open dripproof, 404T
Operating time	3000 h/yr
Energy cost	$.04/kWh

Operating parameters	Motor A	Motor B
Published FL current rating	120 A	118 A
Published voltage rating	460 V	460 V
Actual test FL current	130 A	112 A
Actual test FL voltage	474 V	474 V

Calculated data	Motor A	Motor B
Published efficiency	91.6%	93.1%
Annual savings	—	**$157**

Actual test data	Motor A	Motor B
Kilowatt-hours, initial	77,354	26,427
Kilowatt-hours, final	70,171	20,186
Kilowatt-hours, net	7,183	6,241
Hours, initial	15,963.5	16,140.0
Hours, final	15,842.5	16,009.1
Hours, net	121.0	130.9
Hourly energy usage (kWh)	59.4	47.7
Annual savings	—	$1,404

Comments on results
CASE 1

Voltage variation—Applied voltage is 3% higher than nominal. Because Motor B is designed at a lower point on the saturation curve than Motor A, higher voltage lowers the current draw and increases the efficiency of Motor B. Motor A reacts in an opposite direction.

SIMPLE EQUIPMENT used to test effectiveness of energy-efficient motor included an elapsed-time meter mounted on meter-socket enclosure and a kilowatt-hour meter.

Voltage unbalance of less than 1% results in efficiency loss for both motors, but it could not be determined if there was a significant difference in impact on either motor.

Load variation—This application involved a duty cycle that required the motor to run a length of time unloaded. The no-load losses of Motor A were considerably higher than those of Motor B. No-load energy consumption is an appreciable factor in explaining the difference in average energy consumption recorded on the watthour meters.

CASE 2

Voltage variation—Applied voltage is 4% high for Motor B and 9% high for Motor A. An analysis of the full-load saturation curves for each motor reveal that both motors operate well below the knee of the curve at their nameplate voltage. However, at 600 V, Motor A demonstrated *reduced* efficiency, and Motor B showed efficiency *improvement*.

Voltage unbalance—Both motors were subjected to the same unbalance, which measured less than 1% and did not contribute signficantly to the difference in energy consumption.

Load variation—The driven machine was operated under the same load conditions for each motor, which varied from approximately ¾ to ½ load.

At nameplate voltage, either motor would demonstrate similar drops in efficiency—approximately 1 to 1.5 points. But, under duress of the high voltage applied, Motor A experienced a drop of about 3 points in efficiency at ½ load. This raised its average energy consumption throughout the load range.

CASE 3

Voltage variation—Applied voltage is 2.6% below nameplate nominal. In this application it actually helps the efficiency of both motors, because they are so lightly loaded. Low voltage tends to lower current draw, which correspondingly decreases load losses (I^2R) in the motors.

Voltage unbalance was negligible.

Load variation—Motor A was installed as original equipment and, because it was consistently drawing 60% of its rated amps, it was decided that the next-lower rating (5 hp) should be tested.

As it turns out, a 3-hp rating might have been a better choice for the loads run on the pump at the time. With the

CASE 2. Textile manufacturer

Application Motor type Operating time Energy cost	Yarn finishing 15 hp, 1800 rpm, TETC, 254T 6000 h/yr $.04/kWh	
Operating parameters	**Motor A**	**Motor B**
Published FL current rating Published voltage rating	17.0 A 550 V	14.8 A 575 V
Actual test FL current Actual test FL voltage	14.2 A 600 V	13.0 A 600 V
Calculated data	**Motor A**	**Motor B**
Published efficiency Annual savings	86.7% —	89.8% **$107**
Actual test data	**Motor A**	**Motor B**
Hourly energy usage (kWh) Annual savings	10.1 —	7.7 **$76**

CASE 3. Petro-chemical plant

Application Motor type Operating time Energy cost	Pump 7.5 hp*, 1800 rpm, TEFC, 213T 7800 h/yr $.04/kWh	
Operating parameters	**Motor A**	**Motor B**
Published FL current rating Published voltage rating	11.2 A 460 V	6.1 A 460 V
Actual test FL current Actual test FL voltage	6.8 A 447 V	4.2 A 449 V
Calculated data	**Motor A**	**Motor B**
Published efficiency Annual savings	82.7%** —	87.0% **$70**
Actual test data	**Motor A**	**Motor B**
Hourly energy usage (kWh) Annual savings	3.23 —	2.74 $153

*Replacement motor B was rated 5 hp
**Motor A was calculated from 7.5-hp design performance data at 5-hp load point.

oversized motor, much of the power requirement was the reactive component, which does no useful work output. The energy savings were achieved primarily by sizing the motor closer to the load.

Summary

The three case histories illustrate that energy savings of actual installations often exceed expected savings when initial evaluation uses only nominal data. There were some other cases where installation of energy-efficient motors did not achieve anticipated results. The most probable reason for those cases was either that the standard motor had better efficiency than assumed or the high-efficiency motor had lower. But in the majority of installed tests, the savings realized were better than originally calculated.

The case studies also highlight application variables that directly impact the operating performance of all motors. Variations in power-supply voltage and connected load will cause motors to operate with less than optimal performance. These variables are under the direct control of application engineers and plant-site personnel. By controlling these environmental factors, the user can expect to achieve full benefit from the design improvements available in energy-efficient motors.

Consider practical aspects when buying high-efficiency motors

By RAY E. TIMPONE, Manager of Engineering, Small Motor Div., Marathon Electric Manufacturing Corp., Wausau, WI

JUST A FEW YEARS BACK when a motor needed replacement, the plant electrician would dust off the nameplate, and the normal tendency was to take the easy route—simply order the same brand, size, model, etc. In some instances, it was prudent to check out the motor for possible rebuilding, if appropriate. Likewise, when buying a new motor for a routine application, the plant electrical engineer or chief electrician would often take the sure and simple route—select a brand, a type, hp, speed, etc., that corresponded with an existing installation. This was a fast, practical and effective way to get the job done—back then.

Not anymore. Today, cost of energy is significant, and motors account for the most sizeable chunk of the electric power bill. Therefore, it is vitally important to consider the efficiency of any new or replacement motor installation.

In addition to the usual motor-selection factors—NEMA design, hp, speed, frame size, voltage, enclosure, duty, mounting, etc.—cost must now receive special attention. The initial cost, of course, is important. But now, total owning costs and related factors should also be considered. Furthermore, the total motor installation should be carefully evaluated. This includes not only the motor but also the load and connecting drive train, which could be a direct coupling, belts, gears or other apparatus. The entire motor drive should be carefully selected for maximum efficiency.

Motor-drive efficiency

Many people believe that a motor consumes all the electrical energy that is fed into it. Actually, it consumes only that part that is not passed on to the driven equipment. If 80 kW are delivered to a motor that is 90% efficient, the motor consumes (converts to heat) 8 kW (10% of 80 kW) and transfers the rest of the energy (72 kW) to the pump, compressor, or fan that it is driving. If the pump, compressor, or fan is 60% efficient, then it consumes 28.8 kW (40% of 72 kW) in the act of doing its job. If a pump converts the remaining energy to fluid motion in a pipe, energy is lost in the process. This process continues until each watt can be accounted for. It usually turns out that the motor is one of the most efficient parts of the system.

Frequently, something as simple as alignment between couplings can influence efficiency by a significant amount. As misalignment increases, the energy transmitted through a coupling is reduced. If the cost of energy is $.04/kWh and the efficiency of a 100-hp motor is reduced by 0.5 percentage points, the cost to the user will be in the range of $100-$200 per year or more if the motor runs continuously. Other factors to keep in mind are the importance of good ventilation in the area of the motor, as well as keeping the motor clean. Both will help to maintain efficient operation and extend the life of the motor.

Motor losses

It is important to retain a clear, terse understanding of motor losses and major techniques used to reduce these losses.

The energy that is not transmitted to the driven equipment is usually referred to as the motor losses. These losses are equal to the power input minus the power output. The energy, referred to as the losses, results in heating the various components of the motor.

Motor losses can be broken down into four categories, each with different characteristics.

$I^2 R$ *losses.* These losses result from the energy it takes to drive current through a conductor. This loss is calculated by squaring the current and multiplying by the resistance. There are current-carrying conductors in both the rotor (rotating part) and stator (stationary part), and consequently they both have $I^2 R$ losses. The usual technique to minimize these losses is to install larger copper (not aluminum) conductors in the windings.

Friction and windage. It takes energy to overcome the friction in the bearings. Energy is lost in the bearing and results in raising the temperature of the bearing and bearing grease. Friction adds to the load placed on the motor by the driven equipment. Windage refers to the resistance that air presents to such rotating parts as cooling fans. Cooling fans also add to the load placed on the motor. Smaller fans and antifriction or high-quality bearings will help to minimize these losses.

Core losses. There are two basic components to core losses: hysteresis and eddy currents. A magnetic field is created in the steel cores of the motor. This magnetic field is changing very rapidly. The steel cores have a magnetic memory called hysteresis, which must be overcome. Energy is lost in the process of overcoming this magnetic memory. Hysteresis losses can be reduced by using special, rather expensive steel. Use of these steels depends on the value of the efficiency advantages gained vs the increase in cost.

The other type of core loss, eddy-current loss, results from electric currents that circulate in the steel and produce no torque. Motor manufacturers use steel made from thin sheets and sometimes special materials to minimize these losses.

PROPER COUPLING ALIGNMENT is critical to a motor's overall performance. Dial indicator positioned at motor-load coupling measures both parallel and angular variation. In the typical motor-drive application, losses occur anywhere in the drive train.

Stray load losses. These include several types of losses that are difficult to measure and vary as the load on the motor varies. In some respects it can be said that they are related to motor disymmetrics, such as the existence of rotor and stator slots in the steel cores. Because these losses are hard to quantify, they are generally lumped together and produce much of the variation that is found in different test methods.

An accompanying diagram illustrates the distribution of losses in a typical 25-hp, TEFC motor for various categories. The number in each box shows the percentage of the losses usually encountered in that loss category.

This particular 25-hp motor has an efficiency of 90% at rated hp using IEEE 112A, Method B (see below). If design changes could be made that would reduce the losses by 25%, the gain in efficiency would be only 2.5 percentage points, to 92.5%.

Measuring motor efficiency

Members of the National Electrical Manufacturer's Association (NEMA) have agreed to use a standard created by the Institute of Electrical and Electronic Engineers (IEEE) for testing 3-phase induction motors. The standard is called IEEE 112A—Method B and is important because it provides a consistent efficiency measurement technique for those who use it.

Responsible motor manufacturers feel that this test procedure produces more-accurate results and consequently have accepted it as their test method. Unfortunately, some manufacturers use procedures that tend to result in higher efficiency ratings. Ratings are expressed as "Motor efficiency numbers." These numbers may mislead the user into thinking he is getting more than he really is, unless the proper test has been made.

Also, it is important to know the horsepower delivered by the motor for the given efficiency. Ideally, the motor horsepower and efficiency required will be available at the load in each specific application. It is important to be aware that this horsepower is not necessarily the same as rated hp.

Motor load and efficiency

Efficiency varies as the load on the motor changes. This, of course, makes it important to know what horsepower the motor must deliver throughout its duty cycle to evaluate a motor's energy cost. Efficiency is not always highest at rated horsepower. In fact, in many motors the efficiency is higher at ¾ load than at full load.

A higher-horsepower motor typically has a much flatter efficiency curve than one with a lower horsepower. This means that a 100-hp motor is likely to have the same efficiency at full load as at ¼ rated load, while for a 5-hp motor the efficiency at ¼ rated load will be considerably less than at full load (see accompanying curves).

Voltage unbalance

If the voltage delivered to the motor is unbalanced, a significant amount of energy can be lost. If this unbalance is a result of internal plant loads, perhaps a few modifications will improve the voltage balance and save money. If the service to the plant results in unbalanced voltage, the utility should be consulted and be expected to take corrective action. Sizable reductions in efficiency can occur with unbalance as low as 1% of the rated voltage.

Payback calculations

Much has been written about the various methods of calculating the economics of high-efficiency motors.

All have their place, whether they be simple payback calculations or present-value, life-cycle costing. However, it is also possible to construct "rules of thumb" to judge whether a particular application warrants consideration for a higher-efficiency motor. A good strategy is to concentrate on the number of hours the motor will operate. If you need a replacement motor or a motor for a new application that will be operating for more than 2500 hours per year at rated horsepower, you can count on less than a 3-year payback if you buy a higher-efficiency motor. This assumes $.04 per kWh and a motor that has a significantly higher efficiency than a standard motor. If your ener-

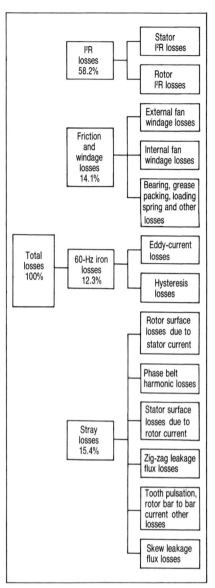

VARIETY OF MOTOR LOSSES for a typical 25-hp, 1800-rpm, TEFC induction motor are shown. The four major categories of losses and typical allocated percentage of losses are given in each box.

gy costs are different, make the appropriate adjustment to the estimated payback. You must also make sure that the motor under consideration really is higher in efficiency. This example assumed that the efficiency difference was two percentage points between the motors being compared.

It would be ideal to make a precise calculation of the life-cycle cost of an electric motor, but sometimes it just isn't practical.

Selecting brushes for motors and generators

By ALFRED O'ROURKE, President, Magna-Wind, Inc.
Electrical-Mechanical Sales & Service, Wallingford, CT

RELIABLE performance of brushed-equipped motors and generators depends on careful choice of the proper brush for the application. Numerous characteristics dictate which brush will be most effective in any given situation. The most-important characteristics are: specific resistance, friction coefficient, current-carrying capacity, maximum operating speed, and abrasiveness.

Specific resistance. Unless otherwise specified, specific resistance is the resistivity of the brush material in ohm-inches (ohms/in. cube) when measured in the length direction of the slab. (Resistance in the direction of width or thickness may be considerably different.) Knowledge of the specific resistance of a brush will help when compared to other brush characteristics during evaluation of a brush application.

Friction coefficient. This is defined as the ratio of the force on a surface to the force required to slide another surface over it. In the case of a brush, the friction coefficient is determined from tests made on a 10-in.-dia slotted-copper commutator with a peripheral speed of 3000 ft/per min. A brush pressure of 4 lbs/in² on brushes having a cross section of 1 sq in. is used for these measurements. Brush friction is influenced by many factors, including brush temperature, brush pressure, current, atmospheric conditions, mechanical conditions, ring or commutator materials, surface films, and speed.

Brush friction increases drastically when brushes operate at temperatures lower than about 60°C and above approximately 100 to 120°C. The resulting high brush friction often causes brush chatter and chipping. The temperature extremes are usually the result of improper electrical loading—either high or low current densities. Thus, it is important to operate brushes at the recommended current density, usually between 35 to 70 A per square in.

Low brush friction is preferred, since brush friction serves no useful purpose. Power required to overcome brush fric-

ROTATING MACHINES include DC motors and generators as well as motors furnished with slip rings. David Kulak, plant superintendent for Magna-Wind, Inc., inspects brushes on reconditioned equipment.

tion is wasted power. One manufacturer broadly classifies brushes as follows:

FRICTION	COEFFICIENT
High	0.40 and above
Medium	0.22 to 0.40
Low	Below 0.22

Current-carrying capacity. The actual current-carrying capacity of a brush is widely influenced by operating conditions such as type of ventilation, continuous or intermittent duty, speed, and other factors. Most manufacturers publish tables recommending amperes/sq in. to be carried by a specific brush. However, this value varies in accordance with the application. Capacity can vary from about 40 A/sq in. for carbon brushes to 180 A/sq in. for metal-graphite brushes.

The current-carrying capacity of a brush depends on the operating temperature. On well-ventilated machines having small brushes with large surface area in proportion to their volume, and where brushes cover only a small percentage of the commutator or ring surface, conventional current densities can usually be doubled without seriously jeopardizing the performance. On

STOCK OF BRUSHES is checked by author Alfred O'Rourke, president of Magna-Wind. The company stocks a wide variety of brushes and specializes in DC rotating machine applications.

the other hand, increasing the current density without making provisions for maintaining a low brush temperature may reduce the brush life many times.

Maximum operating speed. The highest peripheral speed in feet per minute recommended for the collector or commutator on which the brush is to ride is referred to as the maximum speed. The allowable speed depends not only upon the characteristics of the brush material, but also upon the spring pressure, current density, type of brush holder, brush angle, condition of ring or commutator, and atmospheric conditions. Consequently the maximum speed, conventionally listed as a brush characteristic, is only an approximation.

Various types of brushes have been used with success at different speeds and under different conditions. As a result, the maximum speed at which the brush delivers peak performance often is determined by experience.

For example, a dense, strong grade of an electrographitic brush recommended for speeds up to 5000 ft/min also has been used at 12,000 ft/min on turbine-generator rings. Most metal-graphite brushes are listed for a maximum of 5000 ft/min, but they have been used on small generators and other special machines at 8000 ft/min and above.

Abrasiveness. The ability of the brush to prevent excessive buildup of film usually caused by corrosive or oily atmospheres is called the abrasiveness or "polishing action." Abrasive brushes are usually required on flush-mica commutators. The abrasiveness of a brush may be influenced by its hardness, grain structure, and ash content. Brush abrasiveness is classified as low, medium or high. "Low" indicates very little abrasiveness (commonly referred to as "nonabrasive" by the trade), "medium" indices some polishing action, and "high" indicates pronounced polishing action usually obtained by using a material with high ash content or by the addition of a polishing agent.

Brush types

There are four major brush families, classified according to the manufacturing process used and types of carbons and graphites or other materials incorporated into the brush makeup: graphite, carbon and carbon-graphite, electrographitic, and metal graphite.

Graphite brushes are usually made of natural graphite, although some are made of artificial graphite. Natural graphite contains impurities called ash, which give the brushes an abrasive or cleaning action. Artificial graphite is usually purer and less flaky. They are used primarily for special applications such as on steel collector rings on turbo-alternators and in contaminated atmospheres. Many fractional-hp machines also use this type of brush.

Carbon and carbon-graphite brushes are used chiefly where adverse mechanical and atmospheric conditions prevail. Their properties of high hardness, high mechanical strength, and a pronounced cleaning action usually give long brush life under severe operating conditions, although they may not commutate as well as softer grades.

Carbon brushes are usually composed of amorphous materials such as lampblacks and/or cokes. Carbon-graphite brushes commonly consist of mixtures of the amorphous carbons and natural or artificial graphites.

Electrographitic brushes are composed of amorphous carbon materials subjected to exceedingly high temperatures (2400°C and up), changing them physically to a more graphitic structure. Electrographitic brushes usually have higher apparent density, lower strength, lower hardness, and lower specific resistance than nongraphitized brushes of the same initial ingredients. They are very pure and free from abrasive ash. They generally have good commutating characteristics but may not always be used because of high currents, severe mechanical conditions, or atmospheric conditions where pronounced scrubbing action is required to maintain a stable film.

Metal-graphite brushes are generally made from natural graphite and finely divided metal powders. Copper is the most common metallic constituent, but silver, tin, lead, and other metals are sometimes used. Metal content ranges from approximately 10 to 95% by weight. High metal content provides greater current capacity, higher mechanical strength, and also certain combined characteristics of contact drop and friction found only in metal-graphite brushes. This class of brush material is commonly used where high current densities and/or low voltages are involved. Such brushes usually exhibit a definite polishing action. Typical applications are plating generators, battery chargers, welding generators, and other high-current equipment.

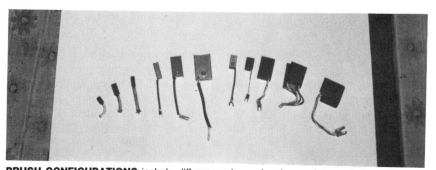

BRUSH CONFIGURATIONS include differences in grade, size, and type of shunt. Shunts, which are usually copper-braided, are attached using a number of methods. They may be riveted, swaged, bolted, or tamped into the brush.

Starting and running induction motors on engine-generator power

MOTORS THAT OPERATE on power supplied from an engine-generator source must be properly selected and applied to assure long life and dependable operation. Of course, the power source must also be chosen with care, since proper voltage and starting capability is vital to the efficiency and reliability of motor operation.

Upon startup, AC induction motors draw several times their rated full-load current. Because of this, a generator-set output voltage will dip until the motor reaches its operating speed. The amount of dip depends on generator preload and generator capacity. The permissible voltage drop depends on the application. Voltage dips up to 30% may be tolerated when starting an unloaded motor; dips greater than 30% should not be permitted because magnetic contactors in the control circuits of other meters on the line may open if the voltage drops below 70% of rated value. Where lighting loads or sensitive equipment share the generator capacity with motors, large or repeated voltage dips usually are objectionable.

Properly matched generator sets operating in parallel will share motor-starting kVA fairly well. Even units of different sizes but of the same manufacturer and configuration furnished with similar items such as generator

type, governor, regulator, etc. will proportionally divide the load according to their respective ratings. It is common practice, or "rule-of-thumb," to increase starting kVA by 10% to account for differences in adjustments, production tolerances, and circulating currents.

The ability to predict the starting kVA division between dissimilar generator sets from different manufacturers is based primarily on experience, and it is advisable to consult with the suppliers of the various equipment involved.

When starting large motors, there are momentary, large kVA demands placed on the generator. Since this is apparent power and not real power, the engine is not affected to a large extent. It becomes critical in starting motors that the generator has the capability to provide this momentary kVA without excessive voltage dip.

The starting kVA of a motor is easily calculated by either of two methods. A NEMA motor code lists the starting kVA per hp using the letters shown in the accompanying chart. The second method is to use the locked-rotor current rating for the motor and calculate

the kVA demand. The locked-rotor current value is standard and is given on the motor nameplate. The calculation procedure is as follows:

$$SkVA = (V \times LRA \times 1.732) \div 1000$$

Where: SkVA = starting kVA
V = supply voltage
LRA = locked-rotor amps

The most-commonly used motor is

AC motor code letters

NEMA code letter	Starting kVA per hp
A	0.00- 3.14
B	3.15- 3.54
C	3.55- 3.99
D	4.00- 4.49
E	4.50- 4.99
F	5.00- 5.59
G	5.60- 6.29
H	6.30- 7.09
J	7.10- 7.99
K	8.00- 8.99
L	9.00- 9.99
M	10.00-11.19
N	11.20-12.49
P	12.50-13.99
R	14.00-15.99
S	16.00-17.99
T	18.00-19.99
U	20.00-22.39
V	22.40-

*Code letters apply to motors up to 200 hp.

the squirrel-cage induction motor, and the most-commonly used type is a NEMA design B, which frequently is classified as a Code F motor. If specific data is not available for any reason, an SkVA of 5.3 kVA/hp can be used for estimating purposes.

When starting several large motors in sequence, it is important to consider the effect that motors already running will have on the generator. The kVA capacity of the generator is limited; therefore, the reserve kVA available becomes more limited as each motor is started. Oversizing the generator relative to engine hp to gain additional motor-starting capability can be advantageous.

In those installations where short voltage recovery time is necessary during motor starting, improved recovery characteristics are obtained when the generator set is at about 75% load, assuming sufficient motor starting kVA is available. At 75% load, the voltage regulator is maintaining a high excitation level.

Where multiple motors are to be started, it is advisable to start the

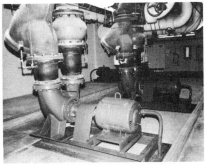

TWO 75-HP INDUCTION MOTORS (above) in boiler room of research facility must operate on power derived from engine-generator sets (right) when normal power is interrupted. Other such motors, as well as critical laboratory loads, are provided with alternate engine-generator supply. After a normal utility-power failure, generator power takes over, and vital loads and motors must be sequenced back into operation appropriately to avoid startup overload of the generator.

largest motors first. Sequencing can be done manually as a part of the startup procedure or with an automatic sequencing device. For engine generator sets, it is important that not all motors on the line attempt to restart at once when the generator set picks up the load. Only motors critical to emergency-power operations should be allowed to restart automatically. If the total starting kVA of these motors plus other emergency load requirements exceed the motor starting kVA rating of the generator set, provision must be made to sequence the starting of these essential motors automatically.

If problems occur during starting of motors, one or more of the following suggestions may be helpful:

1. Specify an oversized generator and improve the system power factor. This reduces the generator set requirement to produce reactive kVA, making more kVA available for starting.

2. Use reduced-voltage starters. This reduces the kVA required to start a given motor. If starting under load, remember that this method of starting also reduces starting torque.

3. Use wound-rotor motors. They require much lower starting current, although they are more expensive than induction motors.

4. Provide clutches so motors may be started before loads are applied to them. While the starting kVA demand of the motor is not reduced, the time interval of high kVA demand is reduced.

5. Change starting sequence; start largest motors first. More kVA is available for starting, although it does not provide better voltage recovery time.

6. Use a motor-generator set. A motor drives the generator which, in turn, supplies power to the motor to be started, as is done, for example, in elevator service. The motor-generator set runs continuously, and the current surge caused by the starting of the equipment motor is absorbed by the motor-generator-set inertia.

Selecting and maintaining explosionproof motors

RELIABLE operation of the thousands of motors supporting research and development work at the huge Hoffmann-La Roche Inc. plant in Nutley, N.J. is the result of careful selection, effective circuit design, and a comprehensive maintenance program.

Throughout this 120-acre, 90-building complex are numerous hazardous areas, most of which are designated Class I, Division 1, Group D locations. Although some areas meet the requirements for a Class I, Division 2 classification, company management has chosen to treat them as Class I, Division 1 areas to simplify the specification of electrical systems and equipment. (Also, a Division 2 location might later be changed to a Division 1 location, and upgrading of equipment would then be necessary.)

Safety must thus be the foremost design consideration. Motor circuits and equipment must meet all applicable NE Code rules, particularly Article 500 (Hazardous Locations), Article 501 (Class I Locations), and Article 430 (Motors, Motor Circuits, and Controllers). In selecting and buying electrical equipment and components, frequent reference to the UL Hazardous Location Equipment Directory (Red Book) supplements use of manufacturers' literature and a common approach backed by years of experience.

Motors for classified areas are selected to meet load and field operating conditions, usually with a TEFC explosionproof, Class I, Division 1 enclosure, dual-rated for either Group C or D, although some motors rated for Class II, Division 1, Groups E and F are in use.

In many of the hazardous areas, motors and processes are controlled automatically from pressurized control rooms. Motor disconnects, controllers, and most other control devices are located in these rooms and thus are not required to be of special design. Positive pressure is maintained by huge blowers; if the pressure drops below a predetermined level, a differential pressure switch sounds an alarm. In these installations, only the motor and an emergency STOP button are installed within the hazardous area. They are, of course, provided with proper explosionproof enclosures.

Controls or other devices required to be installed in hazardous areas are usually provided with a special enclosure designed to be placed under positive pressure as described in NE Code Section 500-1.

Preventive maintenance

Preventive maintenance of explosionproof motors is part of a planned program of regular inspections. Because of the thousands of explosionproof motors installed, only those in critical applications or with a questionable history are included in a regular motor maintenance program. However, most explosionproof motors are checked during a regularly scheduled equipment inspection under an overall preventive maintenance program.

All important or critical equipment or systems are scheduled for periodic inspection by lubricators, mechanics, electricians, and others. During these inspections, the explosionproof motor and other electrical equipment on the machine or system receive a thorough visual inspection.

During scheduled electrical preventive maintenance inspections, the electrician looks for excessive dirt, noise, heat or corrosion. More extensive checks are made during periods when portions of the plant are shut down for maintenance procedures. These checks may include vibration analysis and insulation-resistance tests. All test procedures, instructions and special notes are provided for each major equipment or system. Records of all findings—voltage and current readings, condition of enclosures, belts, clutches, drives, etc.—are recorded for future reference.

Electrical troubleshooting

When an electrical problem in a hazardous area is suspected, either during night-shift inspection or during a regular working day, special procedures must be followed.

First, an experienced plant electrician, trained to work in hazardous areas, is called to observe the problem and make an analysis and judgment based on a visual inspection, if possible. For example, if a motor and/or load appear to be operating improperly, the electrician first visually inspects the components, watching for damaged, discolored or loose parts; listens for any unusual noises; feels the equipment for overheated components; and determines if any unusual odors exist. The equipment operators and area production supervisors often supply clues that help to solve equipment problems.

If further investigation is required, electrical tests can be made easily if control equipment and disconnects are remotely located in a nonhazardous area. If electrical tests must be made

CANNED PUMP is one of many installed in hazardous areas. In this unit, rotor is part of the pump impeller system. Motor stator windings are totally sealed. Connections to motor leads are made via 4-in. round box above unit. See NE Code Section 501-5(f)(3).

HUGE PHARMACEUTICAL PLANT, consisting of nearly 90 buildings on 120-acre site, contains numerous Class I, Division 1 hazardous locations and over 10,000 explosionproof motors.

EXPLOSIONPROOF MOTORS in Class I, Division 1 location are controlled by pushbutton stations in explosionproof enclosures. Note seals in conduit on both sides of pushbutton stations as required by NE Code Section 501-5.

MOTOR CONTROL CENTER is one of many installed in nonhazardous location. Electrical system design specs call for as much electrical equipment as possible to be installed in nonhazardous locations to minimize costs and simplify electrical troubleshooting and maintenance.

15-HP TEFC, Class I, Division 1 motor is provided with flexible, explosionproof conduit and couplings at motor terminal box. Motor leads are sealed in compound where they exit from motor frame and enter terminal box.

EXPLOSIONPROOF CONTROLS include an ON-OFF switch at left, an alarm with pilot light in center, and emergency stop button at right. Note how seals are installed where conduit enters and leaves arcing device per NE Code Section 501-5(a)(1).

LARGE PULL BOX in hazardous area is of explosionproof design. Conduits entering box do not require seals unless the conduits are over 2 in. in dia (see conduit at top of box) as required by NE Code Section 501-5(a)(2).

FLANGE of terminal box on explosionproof motor is carefully cleaned prior to reassembly by Bill Frost, plant electrician. Motor drives agitator kettle via guarded 90° worm gear assembly. RPM indicator is at left.

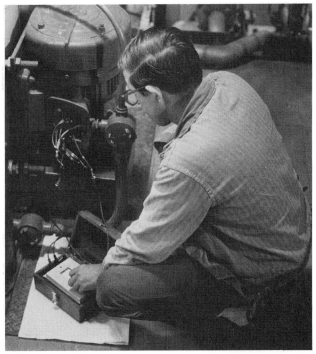

WHEATSTONE BRIDGE is used by John Kelty, plant electrician, to make comparison resistance tests on windings of vertical explosionproof motor. Note explosionproof fittings used to supply power to motor and controls.

within the hazardous area, a special "safety shutdown clearance" must be obtained. This clearance is required even if tests are to be made on de-energized equipment.

The shutdown clearance procedure is carried out by the safety department, area chemists, and specialized technicians. The technicians check temperatures, gas or vapor content of the area and use special instruments, such as an explosimeter, to make sure the area is safe for electrical troubleshooting.

When the safety clearance is given,

the usual troubleshooting procedures are followed. This may include disconnecting the motor from the load to isolate the problem; checking control circuits devices (overload relays, float, limit, pressure switches, etc.); and checking voltage at motor terminals.

The troubleshooting technique includes two special steps:

1. The safest possible troubleshooting tests are selected and used. For example, continuity tests of motor windings would be used prior to a voltage check at the windings.

2. Equipment enclosures are handled with great care during disassembly or reassembly. Mating surfaces are carefully checked for gouges or scratches, since for proper and safe operation, the machined surfaces must be maintained in perfect condition. In most instances, only the terminal box is opened. If further disassembly of the motor is required, a certified service firm is called in.

Accompanying photos provide additional maintenance and troubleshooting details.

Selecting circuits and controls for large air-conditioning machines

HERMETIC REFRIGERANT motor-compressor, rated 2200 amps at 460 volts, is inspected by John Magliano, electrical design engineer, Syska & Hennessy, Inc., and Vito Caputo, chief electrician, New York Maintenance Co., who provide maintenance services on contract for electrical equipment at the building. Five 400MCM THW Cu cables connect to each of six motor terminals.

Here's a step-by-step procedure for applying the NE Code to the design of power circuits supplying 2000-hp, 460-volt cooling equipment for a modern skyscraper.

By ROBERT J. LAWRIE
Associate Editor

SELECTING and designing equipment and circuits for cooling the 45-story Celanese Building in Rockefeller Center, New York, presented difficult-to-answer questions and complex engineering problems. This is because high-horsepower motors, which normally operate at 4160 volts, had to be powered at 460 volts. As often happens on projects such as these, overall design parameters backed by engineering studies showed that a 460-volt supply to the cooling equipment would be most desirable.

Consulting engineers Syska & Hennessy, of New York, determined that this 2-million-sq-ft building would require 7000 tons of air-conditioning equipment and specified electrically driven machinery for the job.

The high-capacity circuits and motors required to handle this large load provide an interesting case study in electrical calculations and design, particularly with respect to NE Code requirements. While the installation was made in compliance with the City of New York Electrical Code, provisions of this code applicable to the air-conditioning portion of the job were in agreement with those of the NE Code, with one exception, which

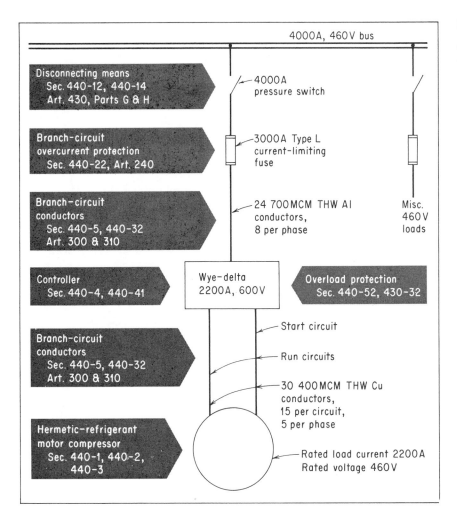

FIG. 1. One-line diagram of motor-compressor branch circuit is typical of four such installations. Circuit data is presented at right of diagram; NE Code references are at left.

will be pointed out. This discussion, therefore, will make reference to requirements of the more widely used NE Code.

Selecting a voltage

The electrical utility brought 13.8 volts into the building and transformed it to 460-volt utilization voltage. The options for selecting a voltage for the motor compressors were, therefore, few.

Use of 13.8 volts would have meant higher operating costs because of the resulting dual metering. The costs of 15-kv switchgear, transformers and maintenance were also prohibitive.

Use of 4160 volts, which usually results in greater economy and highest efficiency for such motors, would have meant stepping up from 460 to 4160 volts. At the capacities involved, initial costs and high costs of operation and maintenance made this plan also unacceptable.

This left 460 volts. Close study showed the job could be done economically and practically at this voltage.

Equipment selected

Syska & Hennessy engineers chose to use four separate air-conditioning units, each powered by a 2200-amp, 460-volt hermetic motor-compressor, and each supplied by a different service entrance for maximum operating reliability. A typical circuit arrangement is shown in Fig. 1.

At each of the four service entrances feeding the air-conditioning load, a 4000-amp bus supplies two 4000-amp fused pressure switches. One supplies building power and lighting loads, the other a motor-compressor.

From the motor-circuit disconnect switch, 24 700MCM, THW Al conductors, 8 per phase are installed in eight 3½-in. aluminum conduit and carry 460-volt power to a wye-delta motor controller. From the controller, 30 500MCM Cu conductors in aluminum conduit carry 460-volt power to the motor.

A wye-delta controller was selected because this type of starter reduces starting current and starting torque to 33% of normal—more than any

HUGE WYE-DELTA MOTOR CONTROLLER reduces motor starting current more than any other type. Start/run contactors (1M) are left; motor-winding contactors (S) are at upper center; run contactors (2M) are at center. Control circuit is in panel section at right. Resistors for closed-transition operation are installed in ventilated cabinets on top of controller.

other type. This was important to assure minimum disturbance to the distribution system during starting of the huge motors. Circuitry requires two sets of leads to the motor—one for use during starting and both for use during normal running. The controller automatically connects the motor windings in wye as the motor is started; after acceleration, the motor is reconnected for normal delta operation. Details of controller operation are explained in Fig. 2.

Applicable code rules

The NE Code in Article 440, "Air-Conditioning and Refrigerating Equipment," provides guidelines for determining sizes and ratings of controllers, disconnects, switches, overcurrent protective devices, and circuit conductors. Section 440-2(a) states that these provisions are "in addition to, or amendatory of, the provisions of Article 430 and other Articles in this Code, which apply except as modified in this Article." Except for Article 430, none of the other articles described (garages, hazardous locations, etc.) apply to the Celanese job. In addition, Articles 422 and 424 are excluded by Section 440-2(b), which states that those articles apply only to refrigeration equipment which does *not* include a hermetic refrigerant motor-compressor.

Because the equipment installed consists only of hermetic refrigerant motor-compressors, Article 440 was followed on this project plus any applicable sections of Article 430 not amended by Article 440.

Marking of equipment

A hermetic refrigerant motor-compressor is defined by Section 440-1 as:

A combination consisting of a compressor and motor, both of which are enclosed in the same housing, with no external shaft or shaft seals, the motor operating in the refrigerant.

Such equipment is required by Section 440-3 to be marked by the manufacturer with certain values under certain conditions in addition to the manufacturer's name, trademark, phases, voltage, frequency, and locked-rotor current. The "rated load current" must appear on all hermetic refrigerant motor-compressors. It is defined in this section as follows:

RATED LOAD CURRENT for a her-

metic refrigerant motor-compressor is the current resulting when the motor-compressor is operated at the rated load, rated voltage, and rated frequency of the equipment it serves.

In addition, if the hermetic motor-compressor incorporates a protection system that permits continuous current in excess of the specified percentage of nameplate rated load current given in Section 440-52(b)(2) or (b)(4), the equipment must be marked with the "branch circuit selection current." This is defined in Section 440-3 as follows:

BRANCH-CIRCUIT SELECTION CURRENT is the value in amperes to be used instead of the rated load current in determining the ratings of motor branch-circuit conductors, disconnecting means, controllers, and branch-circuit short-circuit and ground-fault protective devices wherever the running overload protective device permits a sustained current greater than the specified percentage of the rated load current. The value of branch-circuit selection current will always be greater than the marked rated load current.

The four hermetic motor-compressors supplied on this project were marked with a rated load current of 2200 amps and a locked-rotor current of 9750 amps. They do not have inherent or special protective systems, nor are they marked with a branch-circuit selection current. The 2200-amp value of rated load current, therefore, was used to size circuit components.

Sizing the controller

The size of the wye-delta controller to be used was determined using Section 440-41, which requires the controller to have continuous-duty and locked-rotor ratings not less than the motor-compressor nameplate full load current and locked-rotor ratings. Thus, a 2200-amp, 600-volt controller was ordered, with specs calling for a 9750-amp locked-rotor capability.

Branch-circuit design

The ampacity, type and installation details of the branch-circuit conductors were worked out next. The ampacity was determined on the basis of 2200 amps, the rated load current of the motor-compressor, in accordance with Section 440-5(a).

Conductors between the motor controller and the motor were considered

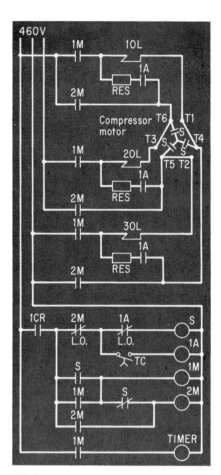

FIG. 2. Wye-delta controller schematic reveals operation of closed transition operation. Resistors in power leads assure that starter is never disconnected from line during start-run steps. This starting method requires two 3-phase circuits from the controller to a six-lead motor. When the start button is pressed, coil 1CR in a pilot circuit is energized, closing contacts 1CR and energizing contactor S. Its contacts close, connecting the compressor motor windings in wye. A normally open auxiliary contact on contactor S closes, energizing contactor 1M, closing its contacts and energizing the compressor motor windings in wye. After a predetermined interval, the normally open contact on timer T closes, energizing contactor 1A. Its contacts close, connecting resistors RES in wye and paralleling them across the wye-connected motor winding. A late-opening auxiliary contact on contactor 1A opens, deenergizing contactor S, opening its contacts, and placing resistors RES in series with the motor winding. The motor is now connected in delta. A normally closed auxiliary contact on contactor S closes, energizing contactor 2M, closing its contacts and thereby shorting out the resistors RES. The delta-connected compressor motor is now energized at full voltage.

first. Because the controller is a wye-delta type starter, two main circuits are required—one serves as a starting circuit, and both are used during normal running of the motor (see Figs. 1 and 3).

The ampacity of these branch-circuit conductors is determined by Section 440-32, which requires branch-circuit conductors supplying a single motor-compressor to have an ampacity not less than 125% of either the motor-compressor rated load current or the branch-circuit selection current, whichever is greater.

This means that these conductors must have an ampacity of at least $1.25 \times$ the rated load current, which is 1.25×2200 or 2750 amps.

However, in this instance, there are two parallel circuits feeding the motor, and when the motor is in normal running mode, the circuits are connected in delta. Because there are two circuits, the required ampacity must be divided between them. But because of the phase-current relationship in delta circuits, the division will not be arithmetic. Fig. 3 illustrates this principle.

The ampacity of each circuit can be calculated by multiplying the rated load current by the constant that applies to current relationships in balanced delta systems and by 125%, as required by Section 440-32:

Rated load current $\times$ K $\times$ 125% = Line current of each circuit.
2200 amps $\times$ 0.58 $\times$ 1.25 = 1595 amps.

Thus, a conductor makeup and arrangement had to be designed for the two circuits between the controller and the motor, each with a capacity of at least 1595 amps. Since this run was relatively short and since the connection at the motor required six terminals (to provide for the interchange of wye and delta internal motor winding connections), they decided to use cable in conduit. THW copper conductors were selected because they offered maximum reliability of terminations, which were subject to motor vibration. Furthermore, the extra flexibility of copper helped simplify connections at the motor terminals.

Because both the controller and motor required six terminals, the conductor makeup had to be a multiple of six to permit an equal number of conductors to be connected to each terminal. After a review of NE Code Table 310-16, which gives ampacities of insulated copper conductors, they found that 400MCM THW conductors would provide the required ampacity and combination of conductors. Ampacity

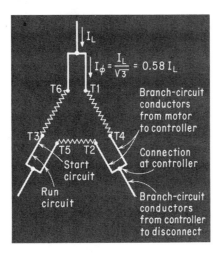

FIG. 3. Delta-circuit diagram shows how branch-circuit conductors between controller and motor connect at corners of motor-winding circuit when motor is in RUN connection. These conductors will carry phase current only, which is 0.58 of the total line current.

for each phase of each circuit was determined as follows:

Ampacity of one 400MCM Cu THW conductor = 335 amps.
5 $\times$ 335 amps = 1675 amps.

This results in a conductor makeup of five 400MCM copper conductors per phase in each circuit, 15 conductors per circuit, or a total of 30 conductors. The conductors are carried in ten 3-in. aluminum conduits, three per conduit. The ampacity of the circuit is the lowest possible with the required circuit arrangement.

The required ampacity of the remainder of the branch-circuit conductors—those between the controller and the branch-circuit protective device, also was 2750 amps, as calculated earlier using Sections 440-5(a) and 440-32. THW aluminum conductors were chosen here because of their lower initial and installation costs. Various cable in conduit makeups were considered to obtain the most economical arrangement. The makeup selected consists of 24 700MCM THW aluminum conductors carried in eight 3½-in. aluminum conduits, 3 per conduit.

Overload protection

Section 440-52 lists four different methods that may be used to protect the motor-compressor, the control apparatus and the branch-circuit conductors from operating overcurrent. Maximum settings of each overload

protective device or system are also given as a specified percentage of nameplate rated load current or branch-circuit selection current.

The four overload protective methods and their maximum rated settings are: (1) separate overload relay—140% of rated-load current; (2) thermal protector integral with motor-compressor—156% of rated-load current (or branch-circuit selection current); (3) fuse or inverse-time circuit breaker—125% of rated-load current; (4) protective system furnished, specified or approved for use with the motor-compressor—156% of rated load current (or branch-circuit selection current). Article 440-21 also requires that provisions of Article 240, "Overcurrent Protection," be applied.

On this installation, overload protection is provided by separate overload relays installed in the wye-delta controller. Trip settings were determined in coordination with the equipment suppliers because of the unusual arrangement of conductors between the controller and motor. Note in Fig. 2 that when the motor is in the normal running connection, all six conductors between the motor and controller carry running current (2200 amps), but that overload relays are installed in only three of these conductors. Actually, strict adherence to Section 440-52(a)(1) would result in the motor-compressor being improperly protected, since the last paragraph in Section 430-32 states:

> Where a separate motor-running overload device is so connected that it does not carry the total current designated on the motor nameplate, such as for wye-delta starting, the proper percentage of nameplate current applying to the selection or setting of the overload device shall be clearly designated on the equipment, or the manufacturer's selection table shall take this into account.

Therefore, the setting of the overload relays was based on the current flowing in *one* of the two circuits to the motor. Because of the delta circuit arrangement, the proper percentage of nameplate current required can be determined by multiplying the constant 0.58 by the rated-load current (2200 amps) to obtain a phase current of 1276 amps. The overload relays were then set at less than 140% of 1276 amps in accordance with Section 440-52(a)(1).

Branch-circuit protection

The required fuse rating or circuit-breaker setting of the motor branch-circuit overcurrent protective device is covered in Section 440-22. This section states that the short-circuit (and ground-fault protective device) shall be capable of carrying the starting current of the motor and shall have a rating or setting not exceeding 175% of the rated load current (or branch-circuit selection current, whichever is greater).

In the event that the protective device cannot carry the starting current of the motor, its rating or setting may be increased, but the rating or setting cannot be higher than 225% of the rated-load current (or branch-circuit selection current, whichever is greater). Thus, the *maximum* rating or setting of the branch-circuit overcurrent device was determined to be 2200 amps × 1.75, or 3850 amps.

Because of the high short-circuit current available, they selected a fusible switch to serve as the branch-circuit disconnect and overcurrent device. This was done primarily because the fusible switch was less costly than a power circuit breaker backed by the current-limiting fuses that would have been required.

Next, they consulted fuse manufacturers' data and, based on motor design criteria, selected a Class L, time-delay, current-limiting fuse rated 3000 amps, 200,000 amps IC. The 3000-amp fuses, which according to fuse data will carry the motor starting current without nuisance tripping, will provide better protection than a fuse rated at the maximum value permitted.

Motor disconnecting means

Section 440-12 states that the disconnecting means serving a hermetic refrigerant motor-compressor must have an ampere rating of at least 115% of the nameplate rated load current or branch-circuit selection current, whichever is greater. Section 440-14 requires that the disconnecting means be located within sight from and readily accessible from the equipment. A small-print note points out that Parts G and H of Article 430 contain additional requirements that must be met.

Section 430-109 in Part H of Article 430 states that the motor disconnecting means shall be a motor-circuit switch rated in horsepower or a circuit breaker. Article 100 defines a motor-circuit switch as:

4000-AMP PRESSURE SWITCH, furnished with 3000-amp, time-delay, current-limiting fuses, serves as motor-compressor disconnecting means. High-capacity load-break-type switch can be operated either manually at the switch or electrically from a remote location.

A switch, rated in horsepower, capable of interrupting the maximum operating overload current of a motor of the same horsepower rating as the switch at rated voltage.

Exception No. 4 of Section 430-109 permits the use of a motor-circuit switch rated in amperes for motors rated over 100 hp.

Because fuses had been chosen for the motor branch-circuit short-circuit protection, the first step in selecting the motor disconnecting means was sizing a fusible switch in compliance with the above code rules:

2200 amps × 1.15 = 2530 amps.

Thus, a 3000-amp switch, furnished with the 3000-amp fuses previously selected, would be suitable.

However, the engineers specified a 4000-amp switch. This was done because the above calculations were made while the motor was being built, and all calculations had to be made on the basis of the manufacturer's motor design data. Using prudent forethought they realized that it was possible, in the event of a design change or unforeseen field conditions, that the actual motor starting or running current, or both, could be higher than the design value submitted. Knowing that the fuse size selected was as low as possible to provide maximum protection (a 3000-amp fuse was chosen based on a maximum permissible value of 3850 amps), they knew it was possible that the 3000-amp fuses might not be able to handle the motor currents drawn during actual operation.

If a 3000-amp switch was installed, it would be impossible to provide fuses of a larger size. Therefore, a 4000-amp switch was specified so that higher rated fuses could be used if necessary. To meet interrupting capacity requirements, they called for a pressure-type switch (see photo).

Disconnect location. Location of system equipment was vital to safe, economical installation. To keep conductor runs as short as possible and to minimize copper losses and voltage drop, all equipment was installed close together and as close as possible to a service entrance. The motor-compressors and controllers were located in an equipment room on a lower level of the building. The motor disconnecting means was installed on the floor above the equipment room. This actually is a violation of Sections 440-14 and 430-102 of the 1975 NE Code, because the disconnect is not within sight and readily accessible from the equipment. However, this arrangement was approved by local code-enforcing authorities because it complied with New York City code rules in effect at the time, which permitted out-of-sight location of the motor disconnecting means if it was capable of being locked in the open position. (This was once permitted also by the NE Code.) If the current NE Code rules had been followed, an additional switch at each controller location would have been required, and this would have added considerable cost.

Arc Electrical Construction Co., Inc., New York, was the electrical contractor.

Motor loading for lowest losses

Given data on induction motor efficiency at any two load points, it is possible to calculate losses at any value of load. These calculations indicate that sizing a motor to operate loaded close to 100% of its rating may not be the most economical way to go.

By JACK WOODHAM
Chief Electrical Engineer
Procter and Gamble Corp.
Cincinnati, Ohio

MOTOR TEST STATION is used for determining losses, power factor and other operating data of actual motor under various load conditions.

SAVING kilowatts has become critical at today's energy costs. Most of the savings available from economical lighting are becoming common practice. One other major electrical source of savings is reducing losses in induction motors. Operating motors at highest efficiency has become important and economically sound engineering. High-efficiency motors are now available at premium prices because reduced losses pay back the higher initial costs. It is important that we be able to determine the losses in any motor at any value of loading if we want to select the most cost-effective motor for a given application.

Determining motor losses at any load

If we obtain, for an induction motor, values of efficiency at two different values of loading, we can determine approximate losses in that motor for any value of loading. Efficiencies are usually available from the manufacturer at 100% and 75% of rated load, so we will use these values.

(Formula 1) $L_{fl} = 0.746 \text{ hp} \left[\dfrac{1}{Eff_{fl}} - 1 \right]$

(Formula 2) $L_{75\%} = 0.746 (0.75 \text{ hp}) \left[\dfrac{1}{Eff_{75\%}} - 1 \right]$

where

L	= losses in kw
0.746	= kw per hp
hp	= rated horsepower of the motor
Eff $_{fl}$	= efficiency at full load
Eff $_{75\%}$	= efficiency at 75% load

Step 1. Determine full-load losses (L_{fl}) and 75% full-load losses ($L_{75\%}$).

Step 2: Losses in a motor are primarily of two types—fixed losses and variable losses. Fixed losses are assumed to be

45

Table 1. Actual vs calculated losses at 50% load

Motor	Rating (hp)	C	Losses at 50% load (kw) Actual	Losses at 50% load (kw) Calculated	% error
A	10	0.4286*	0.7105	0.6940	-2.32
B	50	0.5457	2.0722	2.1384	+3.19
C	30	0.2402	1.2433	1.0698	-13.95
D	50	0.4286*	2.1879	2.1081	-3.65
E	100	0.4286*	3.2435	3.4565	+6.57
F	60	0.3127	1.9461	1.8859	-3.09
G	15	0.2737	0.7218	0.7630	+4.84
H	20	0.3063	0.7831	0.7252	-7.39

*If the efficiencies at the two values of load selected for determining C are the same, the values of C will always be 0.4286. Where actual motor data is not known, for large motors with flat efficiency curves, C may be taken as 0.4286 without introducing a large error.

constant from no load to full load—not exactly true, but close enough so that the errors introduced are not significant. They include core losses, bearing friction, and motor windage.

Variable losses include stray load losses, stator losses, and motor losses, all of which vary approximately as the square of the current (I^2), with the resistance essentially constant. Thus,

$$L_{fl} = (Y_{fl})^2 a + b$$

$$L_{75\%} = (Y_{75\%})^2 a + b$$

where

$Y = \dfrac{\text{output hp}}{\text{rated hp}}$

$Y_{fl} = 1.0$

$Y_{75\%} = 0.75$

a = variable losses (kw)

b = fixed losses (kw)

Substituting the values of Y, these equations become

(Formula 3) $L_{fl} = a + b$

(Formula 4) $L_{75\%} = 0.5625a + b$

Solving for variable losses (a),

$L_{fl} - L_{75\%} = (1 - 0.5625)a = 0.4375a$

$a = \dfrac{L_{fl} - L_{75\%}}{0.4375}$

Solving for fixed losses (b),

$b = L_{fl} - a$

Let us define a new term (c), as the fixed losses per hp of full load.

(Formula 5) $c = \dfrac{b}{L_{fl}}$

We can now develop a formula for the losses at any value of load from no load to full load (L).

(Formula 6) $L = L_{fl} [c + Y^2 (1 - c)]$

Comparing calculated losses to actual losses

To check the results of this formula, let us calculate the losses at various loads for several actual motors using the above method and compare them with data obtained from motor manufacturers. A sample calculation follows, and the results for eight different motors (A through H) are shown in Table 1.

Assume a 10-hp, 1800-rpm, NEMA B motor (motor A in Table 1) has published efficiencies of $Eff_{fl} = 0.86$ and $EFF_{75\%} = 0.86$

(Formula 1) $L_{fl} = 0.746 \times 10 \left[\dfrac{1}{0.86} - 1 \right] = 1.2144$ kw.

(Formula 2) $L_{75\%} = 0.746 \times (0.75 \times 10) \left[\dfrac{1}{0.86} - 1 \right]$

$= 0.9108$ kw.

(Formula 3) $L_{fl} = a + b = 1.2144$ kw.

(Formula 4) $L_{75\%} = 0.5625a + b = 0.9108$ kw.

Subtracting Formula 4 from Formula 3,

$0.4375a = 0.3036$ kw.

Solving for a, b and c,

$a = \dfrac{0.3036}{0.4375} = 0.6939$ kw.

$b = L_{fl} - a = 0.5205$ kw.

$c = \dfrac{b}{L_{fl}} = 0.4286$ (see footnote, Table 1).

The loss equation for this motor at any load then becomes

(Formula 6) $L = 1.2144 [0.4286 + Y^2 (1 - 0.4286)]$

which reduces to

$L = 0.5205 + 0.6939Y^2$

At 50% load (Y = 0.5),

$L = 0.6940$ (calculated).

Manufacturer's data gives efficiency of 84%, so

(Formula 2) $L = 0.746 (0.5 \times 10) \left[\dfrac{1}{0.84} - 1 \right]$

$= 0.7105$ kw (actual)

Error $= \dfrac{0.7105 - 0.6940}{0.7105} \times 100 = 2.32\%$

It can be seen from the data in Table 1 that this method of calculation gives results close to the manufacturers's data. The variation of losses with load depends on the motor manufacturer, horsepower, speed, and frame size. It cannot be predicted, but it can be calculated by the method shown.

Motor efficiencies

Most U.S. manufacturers test the efficiency of motors from 1 hp to 500 hp by the dynamometer method in accordance with IEEE Standard 112-1977, Method B. This gives uniformity to their published data, except that it permits a degree of individual interpretation in applying stray load losses. This results in an undesirably wide range of test results, depending on who does the testing. A new standard, tightening the method of correcting for stray load losses, is being developed to overcome this variation. Japanese motor manufacturers ignore stray losses in determining efficiencies, so for a given motor they would report a higher efficiency than U.S. manufacturers. European manufacturers use still another method. When comparing motors from U.S. and foreign manufacturers, be certain that efficiencies are compared on the same basis.

U.S. manufacturers publish *average* efficiencies for their motors. A manufacturer with close quality control will produce motors close to this average value. A manufacturer with looser quality control may have the same average value, but the efficiency of any individual motor could vary considerably from the average. The errors in the method of loss calculation presented are small considering the variations possible from the average published values for a given manufacturer, and the results can safely be used in making sound management and engineering decisions.

Using the calculations

This method is useful in making comparisons between different-size motors or in choosing between standard and "high-efficiency"motors for a specific load.

EXAMPLE

Assume we want to compare a 40-hp motor with a 50-hp motor with an actual driven load of 37.5 hp. Using the method discussed above (and using efficiencies at full load and 75% load obtained from the manufacturer), we get the following results.

Motor rating	Eff $_{fl}$	Eff $_{75\%}$	c	Losses at 37.5 hp
40 hp	89.0%	88.0%	0.6057	3.512 kw
50 hp	90.5%	90.0%	0.5288	3.108 kw

The actual load is 37.5×0.746 or 27.975 kw, so the efficiency of each motor as loaded is

$\text{Eff}_{40} = \dfrac{\text{kw output}}{\text{kw input}} = \dfrac{27.975}{27.975 + 3.512} \times 100 = 88.85\%.$

$\text{Eff}_{50} = \dfrac{\text{kw output}}{\text{kw input}} = \dfrac{27.975}{27.975 + 3.108} \times 100 = 90.0\%.$

Reduction in losses from the 40-hp motor to the 50-hp motor will be

$3.512 - 3.108 = 0.404$ kw.

Let us also examine power factor and reactive power. From manufacturers' data at 37.5 hp, the 40-hp motor has a power factor of 0.92 at 93.7% load, and the 50-hp motor has a power factor of 0.93 at 75% load. Using the input values determined above (31.487 and 31.083), we can determine the reactive power and kva input by means of the following equations:

kvar = tan θ × kw

Input kva = $\sqrt{\text{kw}^2 + \text{kvar}^2}$

where

kvar = reactive power
kw = input power
θ = angle whose cosine is the power factor.

Summarizing the results of these calculations for the two motors,

	40 hp	50 hp
kw input	31.487	13.083
tan θ	0.426	0.395
kvar	13.413	12.284
kva input	34.225	33.422

Therefore, for the motors used in the example, the 50-hp motor not only uses less energy but also requires less reactive power (kvar). Comparative results of three cases using this method of calculation are shown in Table 2. Figures 1 and 2 show the results of Case 1 and Case 3 in graphic form over a wide range of loads.

Using these results

The conclusion that using an oversized motor will reduce losses is not too surprising. The larger motor has more iron and copper and should have lower losses and operating temperature at a given load. Similarly, oversized transformers are often used to reduce losses and operating temperatures. Older motors were usually built with a "safety factor." More recently, motors have been designed so that operation at rated horsepower was just about their absolute maximum, reducing the use of iron and copper, and resulting in lower initial cost. Then came the rapid rise in energy costs, which seems destined to continue, and initial costs were tempered by costs of operation—losses and efficiency. High-efficiency motors were developed, and one factor in their higher efficiency was more iron and steel. They are again being built essentially oversized, at higher initial cost. The savings in energy can often pay back the higher initial costs in a very short time. In one actual case, a 700-hp motor was chosen over a 600-hp motor for an actual driven load of 560 hp, because the reduced losses produced a payback of the higher initial cost in less than one year.

The method presented for calculating losses at any load for a given motor permits the determination for any loading condition of the kw losses and, given the power factor, the kvar reactive power requirements of various motors. Thus,

Table 2. Motor selection for reduced losses

	Case 1		Case 2		Case 3		
	Motor A	Motor B	Motor A	Motor B	Motor A	Motor B	Motor C
Motor rating (hp)	40	50	20	25	100	125	150
Actual load (hp)	37.5	37.5	18.75	18.75	90	90	90
Efficiency, full load	89.0%	90.5%	89.5%	89.0%	94.0%	93.0%	93.0%
Efficiency, 75% load	88.0%	90.0%	89.0%	90.0%	94.0%	93.0%	93.0%
C (calculated)	0.6057	0.5288	0.5202	0.2554	0.4286	0.4286	0.4286
At actual load:							
Losses (kw)	3.512	3.108	1.685	1.554	5.006	5.087	4.531
Efficiency	88.9%	90.0%	89.2%	90.0%	93.1%	93.0%	93.7%
Power factor	0.92	0.93	0.82	0.86	0.91	0.908	0.89
Reactive power (kvar)	13.413	12.284	10.71	9.02	32.856	33.310	36.708
Total input (kw)	31.487	31.083	15.64	15.54	72.116	72.194	71.654
Total input (kva)	34.225	33.422	18.96	17.97	79.248	79.508	80.510

the differences between motors can be determined, as these motors will actually be operated under load. The more efficient motor will afford savings in energy charges and power factor penalties (or permit smaller power-factor correction capacitors) and have longer life, since it will operate at lower temperatures. However, there will be higher initial motor cost and possibly higher wiring and motor starter costs. These factors can now be evaluated properly for any motor as loaded, permitting the most economical motor to be selected. △

Fig. 1. Variations in motor losses with loading

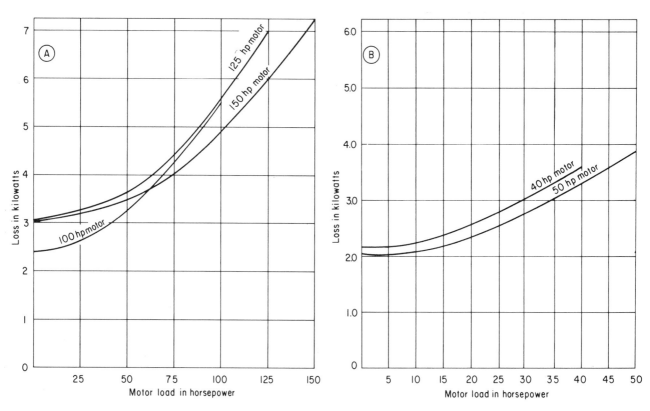

VARIATION OF LOSSES WITH HORSEPOWER is plotted from data calculated by the method described. Note that in Fig. 1A the larger motor has lower losses than the smaller motor at all loadings, but that this does not necessarily hold true for all motors, all loads. As seen in Fig. 1B, the 150-hp motor has lowest losses down to about 65 hp, and the 100-hp motor has lower losses than the 125-hp motor at all loads down to 25 hp, where losses are about equal. Curves were drawn as closely as possible through plotted data points, but at any given load the curve may not pass exactly through the calculated points in order to obtain a smooth curve.

Calculating torque and heat build-up during motor starting

The efficiency of a motor is higher when it operates as close as possible to its full-load rating. However, the T-frame motors now in use have less heat-sink capacity than their predecessors. Knowing torque and heat build-up values during starting will permit motor and load to be matched more closely to the load requirements.

By **WILLIAM C. BRODERICK**
Chief Electrical Engineer
Meyer, Strong and Jones
New York, N.Y.

TFRAME motors are smaller physically than earlier motors of the same horsepower. Less steel is used in the T-frame construction, encouraged by the use of grain-oriented steels in which the molecules tend to point in one direction and not randomly as in ordinary steel. In a very real way, this steel is more efficient.

In addition to more efficient steel, better grades of insulation have been developed. This permits these new motors to operate normally at higher temperatures. These factors, plus the corresponding reduction in copper or aluminum used in the windings, combine to make the T-frame motors smaller. The potential for overheating has increased as a result. There simply is less latitude between operating temperature and maximum permissible motor temperatures.

The hotter a motor runs, and the longer it runs at elevated temperatures, the faster will be its ultimate failure, since insulation is a chemical compound and as such it is time-temperature related. Starting torque and heat generated during starting can be readily calculated; the information gained can then be used to match motor size to load requirements to obtain the maximum energy efficiency for the system.

Information needed to calculate starting torque and heat generated must be obtained from the manufacturers of the motor and driven equipment. This includes the permitted locked-rotor time, speed-torque-current curves of the motor and the load, and the moment of inertia of the load.

Calculations required

The motor must produce enough torque to accelerate its rotor and the load. Whether or not it is capable of doing this can be determined by comparing the speed-torque curve of the motor with the speed-torque characteristic of the load, taking into

Fig. 1 Motor speed-torque-current curves
250 hp, 460 volts, 60 Hz, 3-phase, 1780 rpm
Full-load torque = 740 lb-ft; full-load current = 285 amps; rotor moment of inertia = 31

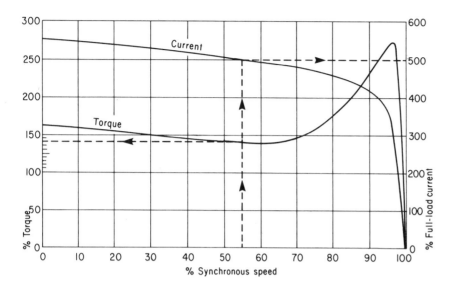

Fig. 2. Forced-draft fan speed-torque curve
1770 rpm; moment of inertia = 1645; inlet vanes closed

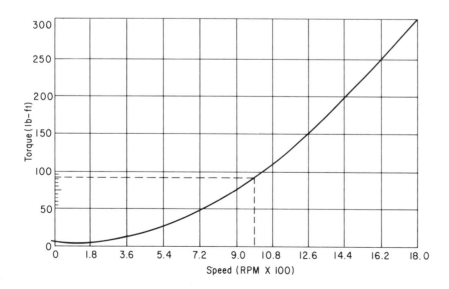

account the moment of inertia of the rotor and the load. Sometimes a simple comparison of the curves will confirm suitability. For larger motors, calculations are usually necessary. The technique to be described here is known as the "graphical integration" method.

EXAMPLE

Figs. 1 and 2, together with data supplied by the manufacturers of the motor and the driven load, will be used to calculate the values shown in Fig. 3.

The curves shown in Fig. 1 are typical for a 250-hp, Design B motor. Fig. 2 is a typical curve for a fan with high inertia characteristics. Assume that the vendors supplied the following data: maximum time permitted to drop from locked-rotor current to operating current, 10 sec; rotor moment of inertia, 31 lb-ft²; load moment of inertia, 1645 lb-ft².

The values shown in columns 2 through 11 of Fig. 3 were calculated as follows.

Column 2. Enter the horizontal scale of Fig. 1 at the center of the speed range as shown by the dashed line. Proceed vertically to the intersection with the current curve and move right, reading the % full-load amperes on the right-hand vertical scale. The dashed lines of Fig. 1 show that the motor will be drawing 500% of full-load amperes at 50 to 60% of synchronous speed.

Column 3. Multiply % full-load amps by full-load amps. For the 50 to 60% speed interval,

$$5.0 \times 285 = 1425 \text{ amps.}$$

Column 4. Again, enter the horizontal scale of Fig. 1 at the middle of the speed interval. Proceed upward to the intersection with the torque curve, move left, and read the % torque on the left-hand vertical scale.

Column 5. The voltage drop at the motor, because of line drop and regulation of the source, is maximum at starting and drops as it accelerates to rated speed. For the motor in question, assume a drop of 20 volts—from 460 to 440 volts—at the instant of starting. This means the voltage at the motor at this instant is 460/440 or 95.7% of rated voltage. For simplicity, the motor voltage during each speed interval is taken as the ratio of the % full-load amps to the % locked-rotor amps multiplied by the initial voltage drop. For the 50 to 60% range, the voltage drop would be 500/560 × 20, or 17.9 volts. (The % locked-rotor amps is obtained from the current curve of Fig. 1 at 0% speed.) The motor voltage at this speed range is thus 460 − 17.9, or 442.1 volts, which is 442.1/460 or 96.1% of rated voltage.

Column 6. Torque varies approximately as the square of the voltage. Therefore, the % torque values are obtained for each speed range by multiplying the value in column 4 by the square of the value in column 5. For the 50 to 60% range, 1.41 × 0.961² = 1.302, or 130.2%.

Column 7. In Fig. 2, the fan manu-

Fig. 3. Results of calculations

1 Motor Speed (%)	2 Motor Current (% F.L.)	3 Motor Current (amps)	4 Motor Torque (%)	5 Motor Voltage (%)	6 Motor Resultant Torque(%)	7 Load Torque (%)	8 Net Torque (%)	9 Net Torque (lb-ft)	10 Elapsed Time (sec)	11 I^2t (amps^2-sec)
0-10	560	1596	160	95.7	146.5	.68	145.8	1078.9	0.91	2,317,967
10-20	550	1568	155	95.7	142.0	1.01	141.0	1043.4	0.94	2,311,107
20-30	535	1525	150	95.8	137.7	2.02	135.7	1004.2	0.98	2,279,113
30-40	525	1496	147	95.9	135.2	4.73	130.5	965.7	1.01	2,260,396
40-50	515	1468	144	96.0	132.7	8.11	124.6	922.0	1.06	2,284,325
50-60	500	1425	141	96.1	130.2	12.50	117.7	871.0	1.12	2,274,300
60-70	490	1397	140	96.2	129.6	18.24	113.4	839.2	1.17	2,283,383
70-80	475	1354	155	96.3	143.7	23.65	120.1	888.7	1.10	2,016,648
80-90	450	1283	195	96.5	181.6	30.40	151.2	1118.9	0.88	1,448,558
90-100	370	1055	270	97.1	254.6	37.50	217.1	1606.5	0.61	678,945
Total elapsed time									9.78	—
Total I^2t									—	20,154,742

facturer's speed-torque curve has been divided into 10 speed ranges of 180 rpm each to match the 10 ranges of the motor curves. Enter Fig. 2 at the middle of the speed range and proceed vertically to the curve, reading the torque on the left-hand scale at the intersection with the curve. For the 50 to 60% motor speed range (which is 990 rpm fan speed), the torque is 92.5 lb-ft. This value is then converted into % *motor* full-load torque. (Later, the final results will be converted from % to lb-ft.) Since the motor full-load torque is 740 lb-ft (Fig. 1), 92.5/740 = .125 or 12.5%.

Column 8. Net torque is the torque developed by the motor minus the torque required for the load: column 6 minus column 7. For the 50 to 60% range, this is 130.2% − 12.5%, or 117.7%.

Column 9. To convert the % net torque to lb-ft, column 8 is multiplied by 740, the motor full-load torque: 1.177 × 740 = 871 lb-ft.

Column 10. The time taken for each speed interval is given by

$$sm/308T,$$

where

s = speed increment
m = moment of inertia (load + rotor).
T = torque (from column 9).

In this example, the speed increment is 180 rpm; the total moment of inertia is 1645 + 31, or 1676. The value 308 is

a constant. Thus, the time spent in the 50 to 60% speed range is (180 × 1676) ÷ (308 × 871.0), which is 1.12 sec.

The increments of time for each speed range in column 10 are then added to find the total time required to accelerate the load to rated speed, in this case 9.78 seconds. Since the motor manufacturer specified a 10-sec maximum time for the motor current to drop to its full-load value, the 9.78-sec value indicates that the motor is suitable for the load from this standpoint.

Column 11. Heat losses in any electrical system are I²R losses; that is, they are determined by resistance and the square of the current. Since both motor rotors and stators have resistance at locked-rotor, heat can be expected to develop. When the mass of the housing, rotor, and conductors are considered in terms of a thermal heat-sink, overtemperature would not seem to be a problem. At starting, however, there is no time to transfer and store heat and reach thermal equilibrium. Thermal limits can be reached in seconds.

The time required to accelerate the load (9.78 sec) in this example is very close to the 10-sec maximum allowed. This could be of concern, especially at initial starting, when the lubrication may be stiff. Most manufacturers will list two values—one for cold starts, one for hot starts.

A relatively simple approximation of I²t may be made to check that the

thermal limitations of the motor will not be exceeded during the staring period. The maximum I²t will be developed with locked-rotor current flowing at the initial instant of starting. From Fig. 1, locked-rotor current is 560% of 285, the full-load amperes, which is 1596 amps. For the 9 sec allowed, the maximum I²t allowed would be (1596)² × 9, or 22,924,944 amp²-sec.

Calculating the I²t for each of the speed ranges of Fig. 3 and totaling them will give the total I²t during the 9.78-sec starting period in this example. This has been done in column 11. For the 50 to 60% speed range, 1425 amps² (column 3) × 1.12 sec (column 10) = 2,274,300 amp²-sec. The sum of all values in column 11 is 20,154,742 amp²-sec, which is less than the maximum permitted value of 22,924,944. It should be noted that this is a conservative figure, since the hot starting time will normally be shorter than the cold starting time calculated and no allowances have been made for heat transfer.

Thus, the motor is suitable for the load and closely matched to the load requirements both from a standpoint of accelerating time and I²t.

For practical, everyday purposes, especially for such low-inertia loads as supply and exhaust fans, a single-step calculation usually is sufficient. Thus, in this example, a single 1800-rpm step could be considered instead of ten 180-rpm steps.

Special ac motors and their applications in industry

Here are four types of ac motors not often seen in industrial plants. Why are they different? What are their operating principles and means of control? What applications do they have?

By ALFRED BERUTTI
Associate Editor

SUBMERSIBLE oil-pump motors, LIMs, PAMs and step motors all represent takeoffs in technology from the standard ac induction motors found in all commercial and industrial locations. Each was developed to answer a particular need not satisfied by the normal T-frame or U-frame induction motor. The theory of their operation, how they are controlled, and the tasks to which they have been applied may suggest new industrial and commercial uses.

Linear motors (LIM)

A linear induction motor differs from the squirrel-cage induction motor in that it converts electrical energy into linear rather than rotational motion. LIMs are at work in things like door operators, power staplers, foil feeders, transportation vehicles, and process equipment mechanisms. Still, they remain as comparatively unknown

items when contrasted with their rotary cousins.

A quick review of basic ac squirrel-cage motor principles is in order. Current flowing in the stator windings magnetizes the stator iron and creates magnetic poles. The magnetic flux from north to south pole completes its circuit through the air gap and rotor iron. This magnetic field is made to rotate by varying the amount and direction of current flow in the stator coils. Faraday discovered that when a conductor forming part of a closed circuit is moved through a magnetic field, current flows in the conductor. The rotating field cutting through the squirrel-cage bars thus causes an induced current. The magnetic field set up by the current flow in the rotor also establishes north and south poles. The north pole of one field attempts to line up with the south pole of the other field. Since the stator magnetic field is

rotating, the rotor field attempts to follow and thereby turns the rotor. When their alignment is prevented by the need to furnish torque to a shaft load, energy conversion takes place. The torque produced is proportional to the angle between the fields. In the squirrel-cage induction motor, slip between stator rotating field speed and the rotor speed provides the means by which lines of force are continually cutting across the squirrel-cage bars, generating voltages in them.

In the linear motor, the stator and its coils are split along its axis and unrolled flat. The magnetic field, instead of rotating, simply sweeps along the length of the stator (termed the "primary" or "motor" in a LIM). The equivalent of the rotor in a LIM is the "secondary" or "armature." The secondary is a conductive material, either copper or aluminum, most often shaped to suit the requirement of the

SECONDARY (ARMATURE) of a LIM is fabricated from a conductive material, usually backed by a steel keeper. The shape is that required for

the application. Steel-backed aluminum disc is used with an open-faced flat LIM, and a copper-plated steel rod is used in a rod-style LIM.

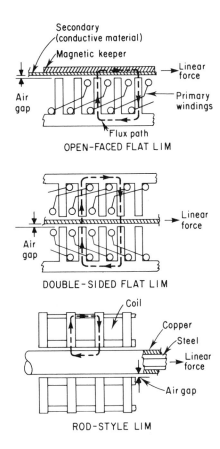

OPEN-FACED FLAT LIM

DOUBLE-SIDED FLAT LIM

ROD-STYLE LIM

SPLIT THE STATOR and the windings of a squirrel-cage induction motor along its axis and unroll them flat and you have a linear motor (LIM). The magnetic field sweeps rather than rotates, producing linear rather than rotational motion. The same principle applies for the open-faced flat LIM, double-sided flat LIM, and the rod-style LIM.

application. As such it can be a rail, a rod, a disc, a sheet of foil, etc. Currents induced in the secondary set up a magnetic field. The interaction of the two magnetic fields produces the required linear force. It is well to note that in the squirrel-cage induction motor, the torque generated acts upon the rotor and stator equally. The torque on the stator is transmitted through the frame of the machine to the foundation and therefore only the rotor turns. In the LIM, it can be either element that is free to move while the other is held in place. This factor increases the flexibility of application of the LIM.

Types of LIMs—There are three basic configurations of LIMs—open-face flat LIM, double-sided flat LIM, and rod-style LIM. The open-face flat LIM follows the configuration description given above. Usually, a magnetic keeper (backing sheet) is provided as part of the secondary conductor. This

acts to provide structural rigidity; but more importantly, it provides a return path for the magnetic flux of the primary. Without the keeper, the heat generated is greater and the power factor of the LIM is lower.

The double-sided flat LIM is simply two open-faced flat LIMs placed opposite each other. One advantage gained by this arrangement is that the iron of the opposed coils acts as the path for the flux, thereby eliminating the need for a magnetic keeper for the secondary.

Take an open-faced flat LIM, roll it up along its length into a cylinder, and you have a rod-type LIM. A soft-steel, copper-coated tube serves as the secondary.

Characteristics and control—Clearance is the key factor to LIM motor performance. The smaller the gap, the greater the efficiency. It is inherently easier to maintain a small uniform gap in an open-faced flat LIM rather than between two opposing faces and the secondary, or between a round rod and the surrounding coils. So, the open-faced flat LIM is not only the simplest; it is also the most efficient.

LIMs are electrically reversible. They provide power in either direction. Be reversing the motor, a braking force can be applied.

As with any electromagnetic device, the limitations of a LIM are set by the amount of heat it generates and the ability it has to dissipate that heat. The iron that separates the coils of the primary also act to dissipate heat; however, in certain applications, additional heat-sinking or auxiliary cooling means are necessary. Another approach for minimizing heat generation is simply by turning off power whenever possible—at the end of a stroke, when coasting, etc. By so doing, the duty cycle of the LIM can be reduced.

Velocity of a LIM is determined by the balance between linear force applied and load characteristics. The LIM accelerated quickly in an attempt to reach its theoretical synchronous speed (160 ips for 3-phase units). Under load, it may reach a speed as high as 90 ips. For many applications, this speed is too high and some form of speed control is necessary. The recommended method of speed control is by using a velocity feedback signal to modulate the power to the motor. A rotary tachometer is used as the feedback source for low-speed applications, while an eddy-current device is used

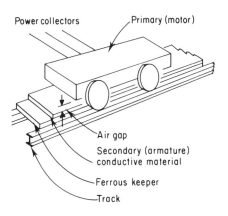

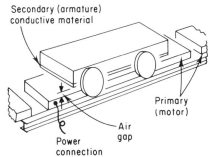

OPEN-FACED FLAT LIM is commonly used for transportation vehicles. Primary and secondary are interchangeable; coils can be mounted on the vehicle or embedded between the tracks. Air cushion or magnetic repulsion can be used in place of track and wheels to maintain air gap.

for higher-speed applications. The signal is used to control the firing angle of an SCR, thus setting the rms voltage to the motor. Since the linear force produced is proportional to power and the square of the applied voltage, force control is achieved, and the LIM maintains constant velocity.

Applications—Possibly the most glamorous use of the open-faced flat LIM is as the prime mover for transportation vehicles. Here, the interchangeability of function of the primary and secondary are most easily visualized. The primary coils can be embedded between tracks. Coil spacing allows the vehicle to coast from one coil to the next. Or, the coils can be mounted on the vehicle and the track or path can be a conductive material backed by ferrous material. The gap between primary and secondary need not depend upon wheels on a rail. An air cushion or magnetic repulsion system would function just as effectively.

Disney World in Orlando, Fla. contains several examples of the use of LIMs for transportation. The Space

Mountain's vehicles, as well as people movers elsewhere in the park, are propelled by LIMs. In these installations, the primary coils are laid between the rails. The plastic-encapsulated coils are 10 in. wide by 15 in. long and 2½ in. thick.

The double-sided flat LIM is most often seen in industrial and commercial locations as door operators. Here the primary is stationary, while the door is attached to the rail-shaped secondary. Gear reducers, clutches and other mechanical devices are eliminated entirely. Since speed can be controlled in both directions separately, the speed of opening and the speed of closing can be set to different values.

Another application for this type of LIM is in foil feeders. The foil, usually aluminum, acts as the secondary. In a typical application, the foil is fed into presses, where it is formed into pie plates, etc.

Rod-style LIMs can be used in any application where a shaft must complete a stroke to move or restrain an object. Strokes up to 10 or 12 ft are possible. Either the rod can be free to act as the prime mover, or the load can be attached to the motor. Rod-style LIMs have been used to feed stock into presses, to provide lifting motion on conveyor transfer applications, as shock absorbing stops (by opposing the weight of traveling objects with a set force), and in place of solenoids for applications requiring motion over 1½ in.

A further extension of the rod-style LIM technology has resulted in a step-rod LIM. In these units, the controller pulses the rod-style LIM. Exact rod positioning can be achieved.

Pole amplitude modulation motors (PAM)

There are several ways to achieve variable flow rates for fans and pumps. The inlet to the system can be throttled, the outlet from the system can be throttled, or the speed may be controlled. By throttling, the output is reduced by the additional pressure drop through the device used to vary the rate. In speed control, both flow and pressure are reduced as a function of driver speed.

Of the methods available for varying flow, the most efficient is speed control. For example: if flow were to be reduced from 100% to 30% of normal, outlet throttling would require 50% of

PAM MOTORS are manufactured from the same constituent parts as standard induction motors. Only an experienced eye could detect that the connection of the stator winding coils are different from the normal pattern.

rated power input by the motor. If inlet control was used, the input power requirement would drop to 30%. With speed control, the input power requirement would drop to only 3%.

While speed control points out a potential for energy savings, it is not always economically viable. But when flow range is wide, or when the drive must operate for long intervals at low flow conditions, speed control should be investigated.

There are several ways to achieve speed control when an ac motor is used as the prime mover. It can be done with a motor having two separate windings, with a consequent pole motor having a single 2:1 ratio winding, with a single-speed squirrel-cage induction motor equipped with an intermediate speed adjusting device, with a variable frequency drive, or with a PAM motor.

Construction—PAM motors are two-speed, single-winding motors. However, they are not limited to the standard 2:1 speed ratios. Ratios can be as high as 10:2 or as close as 6:8.

PAMs are manufactured as standard induction motors. Only an experienced eye could detect that the connection of the stator winding coils varies from that of the single-speed motor. Six leads are brought out to the motor junction box.

Theory of operation—The speed of any motor is determined by:

$$\text{Speed} = \frac{120 \times \text{Hz}}{\text{Poles}}$$

A 4-pole machine will thus have a speed of 1800 rpm at a frequency of 60 Hz. If the windings can be switched so that electrically the number of effective poles can be changed, then speed will also change.

As with the ordinary consequent-pole motor, the windings of the PAM are connected in series at one speed and in parallel at the other. However, by building the motor with the stator coils connected properly, the flux pattern of a group of poles can be superimposed on the original flux pattern. The result is modulation that

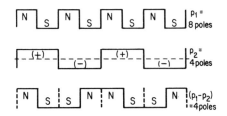

POLE AMPLITUDE MODULATION involves the superimposing (modulation) of one flux pattern (P_2) upon another (P_1) in the air gap of an induction motor. Two values result, ($P_1 - P_2$) and ($P_1 + P_2$). The proper connection of stator windings during manufacture allows one of the resultant pair to be cancelled. A two-speed motor results.

effectively cancels the magnetic effect of some poles, varying the speed.

An analogy to modulation in a PAM is to consider the method used to transmit AM radio signals. The frequency of the spoken word (P_1) is superimposed over (modulates) a second higher carrier frequency (P_2). Two additional frequencies are produced; ($P_1 - P_2$) and ($P_1 + P_2$). Both are high frequencies and can therefore be transmitted over long distances. The modulated carrier frequency is transmitted, and in the receiver the (P_2) signal is filtered out. Only the voice frequencies remain. It is the ($P_1 - P_2$) and ($P_1 + P_2$) values that are produced by superimposing flux patterns in the PAM that is used to achieve the second speed. The connection pattern of the windings allow either ($P_1 - P_2$) or ($P_1 + P_2$) to be eliminated, thus achieving two speeds at any ratios desired.

Starting and switching—Starting and switching can be accomplished using a standard starter plus a 5-pole auxiliary contactor. More often it is performed by using a starter in combination with a motorized switch. When changing speeds, the motor is momentarily deenergized by opening the starter. After a short delay the transfer of switch position is made, and the starter is energized.

Applications—It is common practice to run forced-draft and induced-draft fans at less than their maximum capacity for extended periods. Two-speed motors allow for such cyclic operation as well as the case where provisions must be made for future capacity.

Retrofitting of existing equipment to achieve maximum energy efficiency is an increasingly important consideration. By matching the need to the speed as closely as possible, energy can be saved.

For those applications where lower-than-normal motor inrush current is required, PAMs can be used. By starting on the low-speed winding and then transfering to the high-speed winding, a reduction in inrush can be accomplished. This offers an alternative to reduced voltage starting methods.

PAMs are also available as 2-winding, 4-speed motors and in all three categories of multispeed application—variable torque, constant torque, and constant horsepower.

Step motors

A step motor is an electromagnetic device that converts digital electrical signals into fixed mechanical movement. Conventional motors rotate continuously when energized. A step motor rotates in fixed angular increments each time it is energized. The step angle is fixed and is determined by the construction of the motor. Commonly available step motors range from step angles of 0.72 degrees (500 steps per revolution) to 45 degrees (8 steps per revolution).

How they work—In the majority of applications of step motors, a permanent-magnet type motor is used. In a simplified form, this motor is made up of a stator with teeth that are magnetized to form north and south poles when dc power is applied to the appropriate stator windings, and a two-part rotor separated by a permanent magnet. The rotor is also toothed; but the south teeth are offset from the north set so that when a rotor S is attracted to a stator N, there is a rotor N that is simultaneously attracted to a stator S. By selectively applying dc voltage to the stator windings, the stator N and S poles can be caused to rotate. The rotor will step until the closest N and S of the rotor line up with the S and N of the stator. The discrete steps are achieved by having an unequal number of teeth in the stator and the rotor. For example, if there are four stator poles, then there would be five rotor teeth. Therefore, only one set of teeth can line up at one time. The exact number of teeth determines the step angle.

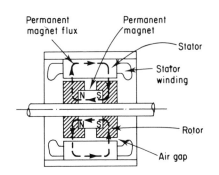

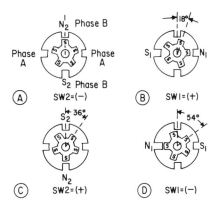

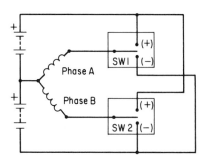

Step	CW rotation		CCW rotation	
	SW 1	SW 2	SW 1	SW 2
1	Off	(−)	Off	(−)
2	(+)	Off	(−)	Off
3	Off	(+)	Off	(+)
4	(−)	Off	(+)	Off
1	Off	(−)	Off	(−)

BY SELECTIVELY CLOSING switch 1 or 2 and applying dc power to the stator winding of a dc step motor, the north pole can be made to rotate. The nearest south pole of the rotor is attracted, causing indexing of the rotor in the desired direction.

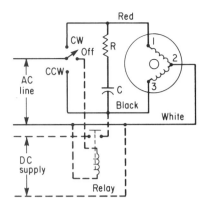

CONNECTED TO A 120-VOLT ac source, a properly wound permanent-magnet step motor becomes a constant-speed synchronous motor. A phase-shifting RC network must be used to convert the single-phase excitation to the required two-phase. A dc power source can be added to energize one winding when the motor is stopped, thereby increasing the holding torque.

IN INDUSTRIAL LOCATIONS ac synchronous-inductor motors are often used for accurately adjusting gaps between rolls in coating operations.

The control of a step motor involves sequential switching of dc power to the stator windings. Depending upon the controller used to supply the pulses, stepping rates of 10,000 full steps (or 20,000 half steps) per *second* can be achieved. Typically, the controller would interface with a digital controller or computer for precise positioning of machine-tool components.

At this point, the question may arise, "What is a discussion of dc step motors doing in an article on ac motors?" The answer is that the permanent magnet dc step motor was originally developed as a slow-speed synchronous inductor motor! An appropriately wound dc permanent-magnet motor, connected to a single-phase, 120-volt, ac supply, can be operated as an ac synchronous, constant-speed motor. All that is necessary to be added is an RC network for phase shifting (not required if operated from 2-phase source), and a 3-position switch to provide FORWARD-OFF-REVERSE control. Virtually instantaneous starting and stopping characteristics make the ac version ideal for precise motion control. Generally, the motor will start within 1½ cycles and stop within 5 mechanical degrees. Full synchronous speed is reached within 5 to 25 milliseconds. No braking system is required to achieve the small stopping distance; simply de-energize the motor.

Current flows through the stator windings only when the motor is energized. No current flows through the rotor. Thus, ac starting and running currents are virtually identical—there are no high inrush currents.

The torque output of the motor is proportional to the applied voltage, and the speed of the motor is directly proportional to the applied frequency. Therefore, within limits, both torque and speed are adjustable.

Once the motor has been deenergized, residual holding torque is supplied by the permanent-magnet rotor. If more holding torque is required, dc power can be applied to one or both of the stator windings. Holding torque of the two windings will be increased by 120% and 150% above the rated residual torque of the motor.

Applications—The ac synchronous version of the dc step motor is most often found in industrial environments, applied where precise motion control is necessary. Gaps of coating rolls in high-speed treating lines must frequently be adjusted to apply precise thicknesses of finish to materials. The permanent-magnet synchronous motor has been used to change and maintain these gaps. They are also seen in edge-guide controls and as remote controls for switches and rheostats.

Submersible oil-well pump motors

We often think of oil wells in terms of pictures we have seen of great torrents of oil spewing from the ground when a well has been brought in. The gusher covers drilling rig, crew, and

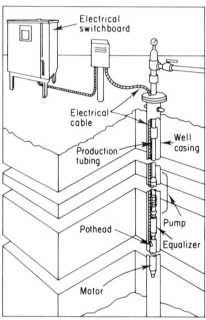

SUBMERSIBLE oil-well pumps are made up of the motor, equalizer, and pump. In a typical installation, the unit is suspended on the production tubing by threading the tubing directly into the pump discharge head. The direction of rotation is such that all threaded joints tend to tighten during operation. A special cable connects the motor to the electrical switchgear.

landscape with "black gold" until it is capped off. Actually, subsurface pressure most often is not sufficient to keep oil flowing to the surface of a producing well. One of the means used to achieve artificial lift is with an electrical centrifugal subsurface (downhole) pumping system.

SWITCHGEAR for oil-well pump contains a disconnect means, contactor, overload relay, control and monitoring equipment. Standard voltage ratings are 440, 880, 1500, and 2500 volts.

How can a pump and its motor be lowered thousands of feet into a well when the casing of the well has an outside diameter of only 5½ in., 7 in., or 8⅝ in.? The answer is by making the entire motor-pump assembly only 4⅘ in. in diameter—for a 5½-in. OD pipe—and up to 120 ft long! Obviously, this is not the standard NEMA frame-size, squirrel-cage induction motor that we are dealing with.

Motor design—Motors used in submersible oil-well pump service are usually 3-phase, 3500-rpm induction motors. Ratings range up to 275 hp for 5½-in. OD casings, and up to 550 hp for 8⅝-in. OD casing wells. The motors are oil-filled and are cooled by the transfer of heat to the well fluid moving past the motor and into the pump. The pump is located above the motor in the typical configuration.

Temperature is an important factor at the depths at which submersible oil-well pumps are used. The farther down, the higher the temperature. Thus, while the ambient temperature is considered to be 40C for the standard application of a squirrel-cage induction motor, submersible oil-well pump motors are rated to carry full-load at ambients of 120C. Also, insulation systems have been developed that exceed the total allowable temperature for Class H insulation.

The motor itself can be up to 30 ft in length for a 275-hp motor used in a 5½-in. pipe application. Of this length, approximately 29 ft are used up for the stator iron. Bearings are inserted approximately every 14 in. within the stator to support the rotor. A built-in oil filter continuously cleans and filters the motor oil to insure a constant supply of clean oil to the bearings. Otherwise, the rotor and stator are both of standard squirrel-cage construction.

Motor leads are brought out through a pothead fitting. This insures good sealing around the cable and prevents contamination of the motor oil by the well fluid. As an additional safeguard against introduction of well fluid into the motor, a unit called an equalizer is mounted between the motor and the pump. The equalizer serves as a barrier to the migration of fluid along the shaft connecting the pump and motor. It also allows for free expansion and contraction of the motor oil without danger of contamination.

Electrical distribution system—If the configuration of the submersible oil-well pump motor seems unusual, then consider the voltage ratings. The standard voltages for these pump motors are 440 and 762 volts. Higher voltages up to 2300 volts are also available.

For a given wire size and voltage drop, a load can be located three times as far away when fed at 762 volts than if it were fed at 440 volts. In applications where downhole distances run into thousands of feet, the advantage of the higher voltage level is obvious. The rule-of-thumb used to set depth limits is: do not consider a 762-volt motor if the horsepower x the depth exceeds 200,000. For greater depths, higher voltages should be used.

Why was a voltage level of 762 volts made a standard for submersible oil well pump motors? Because there are several convenient ways of obtaining this voltage at an installation that already operates on 440 volts. The first is by the use of autotransformers connected to the existing 440-volt system. The second is to use three single-phase, standard, 440-volt transformers. By connecting their secondaries in wye, 762 volts is obtained phase-to-phase.

Specially designed 3-kv cable is normally used from starting equipment to motor. The cable can be flat or round, armored or unarmored. The ampacity of the cable varies from NE Code values because of the unusual thermal environment in which it must operate. The maximum ampere limits are somewhat flexible, since the temperature of wells can vary considerably. A "cold well" may run from 75F to 125F. "Hot wells" may exceed 250F.

Voltage drop is an important consideration in the selection of the cable size. With the distances experienced in a typical well application, voltage drop is often the limiting factor. By oversizing a cable, voltage drop can be reduced. However, space limitations in the well casing limit this option. The next alternative is to boost the surface input voltage. This is most often accomplished by taking advantage of the transformer taps. In a 762-volt system, the two 2½% taps allow the voltage to be boosted to 800 volts.

Starting equipment and controls—Switchboards are used to house the starting equipment, monitoring devices, and control equipment. Generally they are NEMA 3 weatherproof. The starters used are those that are normally used for across-the-line starting of the standard squirrel-cage induction motor. This includes a disconnect device, contactor, overload relay, and 440/110-volt control transformer. Recording ammeter, undercurrent shutdown relaying, and lightning arresters are also usually mounted in the enclosure.

What sets submersible oil-well pump switchboards apart from the starting equipment found in most commercial and industrial sites are the voltage ratings. The standard ratings are 440 volts, 880 volts, 1500 volts, and 2500 volts. These voltages represent the maximum voltages that are likely to be encountered on the various voltage distribution systems.

Applications—Besides service as submersible pumps in oil wells, these pumps have been used in a variety of different applications. They are used on off-shore drilling rigs as sump pumps, fresh-water supply pumps, cooling-system pumps, fire-protection pumps, and crude-oil transfer pumps. They find application as storage-cavern and mine-shaft evacuation pumps. Submersible oil well pumps can also act simultaneously as a production and injecting pump. By providing lift while injecting the fluid into a reservoir at required presures, surface injection pumps and structures can be eliminated.

PART II

INSTALLATION METHODS

Guidelines for

Effective motor installation

Here are latest techniques for installation, connecting, and startup of modern motors—covering location, mounting, couplings, terminations, testing, and NEC rules.

PROPER installation of an electric motor is essential to obtain top-quality operation, efficient performance, and maximum reliability. For a totally cost-effective installation, procedures should consider all aspects of

By ROBERT J. LAWRIE, Associate Editor

engineering, design, selection, application, and maintenance as well as the details of assembly, hardware, and interrelationship of components and materials. The work demands close coordination, planning, and teamwork on the part of the engineers, installers, and those responsible for maintenance;

and completion of a successful motor installation requires that the latest and best construction techniques be employed.

Receiving and handling

When a motor is received, it should be thoroughly inspected for scratches, dents, or other signs of damage. This inspection should be done before the motor is moved from the shipper's truck or other vehicle. Examine all literature provided with the motor. Do not remove tags pertaining to assembly, storage, lubrication and operation. File all literature with specifications and drawings pertaining to the motor for reference during installation and for guidance during startup and operation.

Handling of large, heavy motors should be supervised by qualified riggers with experience in the use of cranes, hoists, jacks, rollers, wire ropes, cables, hooks, slings and other equipment that may be needed.

Always check the motor nameplate for proper voltage, phase, frequency, horsepower, etc. Large motors are sometimes shipped disassembled. When assembling, be sure all mating parts are clean. Cleaning can be done with a magnet, vacuum cleaner, or dry compressed air (air pressure less than 60 psi).

On smaller motors, turn the shaft by hand to be sure that it turns freely. If the motor is equipped with antifriction bearings, they will normally be prelubricated and ready for operation. However, large motors having sleeve bearings are usually shipped *without* lubricating oil in the bearings; often they are filled with an antirust fluid. Bearings should be inspected through the sight-glass and bearing-drain openings for any accumulations of moisture and

MOTOR INSTALLATION includes replacement of older design motors with new, energy efficient types such as the 50-hp induction motor being installed to boost efficiency of cooling equipment required by a computer room. Proper handling, mounting, coupling, alignment, foundations and terminations are an essential part of the job.

HAZARDOUS AREA in a refinery is a Class I, Group D, Div. 2 location. All four motors, which are *not* required to be explosion-proof, are driving pumps. In foreground are two TEFC units; the two motors in the background are furnished with weather-protected, Type 1 enclosures. Note concrete foundations, steel base supports, and conductors grounding motor frames to steel.

traces of oxidation. Then, fill the bearing reservoirs to normal level with a high-grade industrial lubricating oil.

Safety procedures

Safety is of paramount importance during the installation, startup and operation of motors. Safety starts with proper design, application and selection of the motor and associated components. Be sure that the motor has been well matched to handle the type of load to be driven. Be certain that the enclosure is suited to the surrounding environment and that there is adequate ventilation to assure operation at or below motor design temperature. Check that the motor, gears, belts, driven machinery, etc. are guarded so that anyone near the installation will not be harmed by accidental contact.

All personnel involved with the installation should be familiar with NEMA MG2, *Safety Standards for Construction and Guide for Selection, Installation and Use of Electric Motors and Generators.* Pertinent NEC rules, especially Article 430, and all local safety rules must be observed. In addition, OSHA rules must be studied and followed during the installation of motors and controls. These regulations are included in Part 1910 of the *Occupational Safety and Health Standards.* Obtain a copy of this document from any local OSHA office.

Location considerations

An open-type motor is usually the best choice for installation in surroundings reasonably free of moisture, dust or lint. Be sure space is available for maintenance or replacement. Open motors having commutators or collector rings must be located or protected so that sparks cannot reach adjacent combustible material. This does not preclude the mounting of such motors on wooden platforms or floors.

Dripproof motors are intended for use where the atmosphere is relatively clean, dry and noncorrosive. Keep windings clean with a soft brush, cloth or suction. Totally enclosed motors may be installed where dirt, moisture and corrosion are present, or in outdoor locations. If a drain plug is provided in the end bracket or bell, it should be removed periodically to drain any accumulated condensation. The motor should be installed to deliver adequate power safely. Temperature rise of a standard motor is based on operation at an altitude not higher than 3300 ft above sea level.

When unusual environments or conditions exist (high temperatures, extreme vibration, etc.), special enclosures or arrangements must be incorporated into the installation.

Always try to locate the motor in the best possible environment—a clean, dry, cool location. Often, electrical equipment rooms are specially constructed so that motors and other equipment will operate in a suitable environment and obtain long life and simplified maintenance. However, many motor enclosures are available that will permit operation in a variety of environments. The specific application and the nature of any contaminants present will dictate the best enclosure for the job.

Moisture problems require special consideration. Suitable guards or enclosures must be provided to protect exposed current-carrying parts of motors and the insulation of motor leads where dripping or spraying oil, water, or other injurious liquid may occur, unless the motor is specially designed for the existing conditions.

In addition to open (general-purpose), dripproof, and totally enclosed fan cooled (TEFC), a number of other designs are available for specific environments and applications. For standy service or for damp-location operation, a low single-phase voltage (on the order of 5 to 10% of rated voltage) is sometimes applied to the windings to combat moisture. Some larger motors are available with built-in strip- or tubular-type space heaters for this purpose.

Foundations

A rigid foundation is essential for minimum vibration and proper alignment between motor and load. Concrete, reinforced as necessary or required, makes the best foundation, particularly for large motors and driven loads. In sufficient mass, it provides rigid support that minimizes deflection and vibration. It may be located on soil, structural steel, or building floors, provided the total weight (motor, driven unit, foundation) does not exceed the allowable bearing load of the support. Allowable bearing loads of structural steel and floors can be obtained from engineering handbooks; building codes

TYPICAL MOTOR INSTALLATION on roof of a large industrial plant consists of a 125-hp, 460-V, TEFC induction motor driving a large induced-draft fan for a boiler. The motor enclosure is designed for outdoor service. Large concrete base and heavy steel beams and plates provide solid support for the motor. Note dual lifting hooks used for positioning the 2600-lb motor.

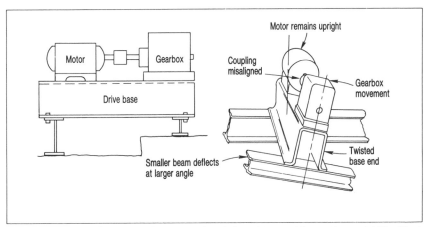

MOTOR MOUNTING on steel base must be designed to provide maximum rigidity. Steel supports should be of adequate size and strength and should be braced against twisting, which can occur during operation of large, high-torque motor drives.

of local communities give the recommended allowable bearing loads for different types of soil. For rough calculation, the subfoundation weight should be approximately 2½ times total weight supported.

In the event that the motor must be mounted on steel, all supports must be of adequate size and strength and braced to assure maximum rigidity.

Whether the motor base is concrete or steel, it must be level. If concrete, be sure it is not too high. A motor can always be raised by use of shims; but

reduction of height by removal of some of the concrete surface would be difficult.

The requirement for a level base is critical. Usually, for a motor installation, there will be four points of mounting—one at each corner of the mounting base. Then there will be mounting requirements for the driven load. All mounting points must be on the exact same plane, or the equipment will not be level.

Before pouring the concrete foundation, locate foundation bolts by use of a template and provide secure anchorage (not rigid). It is recommended that a fabricated steel base be used between motor feet and foundation. See certified drawings of motor, base, and driven unit for exact location of foundation bolts.

Mounting

For smaller motors, sliding bases and adapters are available for use with T-frame motors when they replace an old motor. Also, check whether other components or equipment, such as gears, special couplings, and pumps are to be mounted on the motor. If so, be sure space is available.

After the motor base is in place and before it is fastened, shim as required to level. Use spirit level (check two directions at 90°) to insure that motor feet will be in one plane (base not warped) when base bolts are tightened. Set motor on the base, install nuts, and tighten. Do *not* make a final tightening until after alignment. NEMA standards give dimensions of foot mountings and some flange mountings. Experience has shown that any base-mounted assemblies of motor and driven units temporarily aligned at the factory, no matter how rugged or deep in section, may twist during shipment. Therefore, alignment must be checked after mounting.

Drives

Direct-connected motors with ball or roller bearings may be coupled to the load through flexible couplings. A coupling should not be installed by hammering or pressing. Always heat the coupling to install it on the shaft. Accurate mechanical lineup is essential for successful operation. Mechanical vibration and roughness during the operation of the motor may be indications of poor alignment. In general, using a straight-edge across and a feeler-gauge between coupling halves is not sufficient. It is recommended that the lineup be checked with a dial indicator and checking bars connected to the motor and load machine shafts.

Sleeve bearings are supplied with a babbited face to restrain axial rotor movement during startup or while running disconnected from the load. These babbited faces are not intended to withstand continuous thrust loads, and care must be exercised in the lineup to prevent this from occurring during operation. Lineup should provide operation in the *approximate mechanical*

SLIDING BASES made of steel channel permit motors to be aligned with belt-driven loads. Motors, which range in size from 5 to 20 hp at 460 V, drive air-powered conveyors. Note that driven loads are also provided with adjustable steel bases, which will simplify alignment.

center between the extremes of end-play; this is very close to the magnetic-center location. Standard motors are supplied with more end-play. It is necessary that a limited end-float flexible coupling be used on sleeve-bearing motors to limit the total axial movement to less than that shown in the motor outline drawings. As noted in NEMA Standard MG1-14.38, sufficient thrust to damage bearings may be transmitted to the motor bearing through a flexible coupling.

Ball-bearing motors, unlike sleeve-bearing motors, should be coupled to provide more end-play in the coupling than in the motor. This is because ball bearings will take enough thrust, without damage, to slip the coupling axially to accommodate thermal expansion in the system. The end-play of these motors may be as much as 50 to 150 mils, and the coupling should have at least this much float. Correct axial positioning can be obtained by tilting the motor toward the outboard end moving the rotor as far as it will go in that direction (the rotor will not be easy to move axially, since bearings must slide in the housing) or by barring the rotor over to the outboard end, and then positioning the motor to give at least 150 mils between the coupling

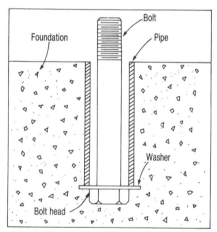

FOUNDATION BOLTS installed within a steel pipe embedded in the concrete foundation provides some freedom of movement of the bolts in aligning them with holes in the motor base.

halves or shaft ends. Or, the motor may be positioned without regard to rotor position so that coupling will allow 100 mils travel in either direction.

Belt drives require that the motor be mounted on slide rails or bedplate with provisions for adjusting the belt tension. Align the pulleys so that the belts run true (perpendicular to the shaft) and with uniform tension on all belts.

NEMA designations for motors

OPEN

General purpose. Ventilating openings permit passage of external cooling air over and around the windings of the machine.

Drip-proof. Ventilating openings are so constructed that successful operation is not interfered with when drops of liquid or solid particles strike or enter the enclosure at any angle from 0 to 15 deg downward from the vertical.

Splash-proof. Ventilating openings are so constructed that successful operation is not interfered with when drops of liquid or solid particles strike or enter the enclosure at any angle not greater than 100 deg downward from the vertical.

Guarded. Openings giving direct access to live or rotating parts (except smooth shafts) are limited as to size by the design of the structural parts or by screens, grills, expanded metal, etc., to prevent accidental contact with such parts.

Semi-guarded. Part of the ventilating openings, usually in top half, are guarded as in a "guarded machine," but others are left open.

Drip-proof fully guarded: A drip-proof machine with ventilating openings as in a "guarded" machine.

Externally ventilated. A machine ventilated by means of a separate motor-driven blower mounted on machine enclosure. Mechanical protection may be as defined above.

Pipe ventilated. Openings for admission of ventilating air are so arranged that inlet ducts or pipes can be connected to them.

Weather-protected. *Type I:* Ventilation passage are so designed as to minimize entrance of rain, snow and airborne particles to the electrical parts. *Type II:* Ventilating passages at intake and discharge are so arranged that high-velocity air and airborne particles blown into the machine by storms or high winds can be discharged without entering the internal ventilating passages leading directly to the electrical parts.

ENCAPSULATED / SEALED

Encapsulated windings. An alternating-current, squirrel-cage machine having random windings filled with an insulating resin that also forms a protective coating.

Sealed windings. An alternating-current, squirrel-cage ma- chine making use of form-wound coils and having an insulation system which, through the use of materials, processes, or a combination of materials and processes, results in a sealing of the windings and connections against contaminants.

TOTALLY ENCLOSED

Nonventilated. Enclosure prevents free exchange of air between inside and outside of case, but not airtight, not equipped for cooling by external means.

Fan-cooled. Equipped for exterior cooling by means of a fan or fans, integral with machine, but external to the enclosing parts.

Fan-cooled guarded. All openings giving direct access to fan are limited in size by design of structural parts or by screens, grilles, expanded metal, etc., to prevent accidental contact with fan.

Explosion-proof. Designed and built to withstand an explosion of gas or vapor within and to prevent ignition of gas or vapor surrounding machine by sparks, flashes, or explosions that may occur within machine casing.

Dust-ignition-proof. Designed and built to exclude ignitible amounts of dusts or amounts affecting performance or rating to prevent ignition of exterior dust on or in vicinity of enclosure.

Waterproof. Designed to exclude water applied in a stream from a hose, except that leakage is permitted around the shaft, provided that water is excluded from oil reservoir and means of drainage is provided.

Pipe-ventilated. Openings so arranged that inlet and outlet ducts or pipes may be connected to them for the admissions and discharge of ventilating air.

Water-cooled. Cooled by circulating water; the water or water conductors came in direct contact with the machine parts.

Water-air-cooled. Cooled by circulating air which, in turn, is cooled by circulating water.

Air-to-air-cooled. Cooled by circulating internal air through heat exchanger which, in turn, is cooled by circulating external air.

See NEMA Standards MG1-1.25, 26, 27 for additional details.

The slide rails should be located so that the motor is near the end of the slide rail closest to the driven machine. This permits maximum adjustment (travel) for belt tensioning and readjustment later to compensate for belt wear or belt stretch.

Tighten the belts just enough to prevent slippage at the rated horsepower. Excessive belt tension causes unnecessary loads on the shaft and bearings. On high-inertia loads or equipment that could jam or stall—where the belt squeal or slip during acceleration or where the torque approaches pull-out torque during overload or stall—tight- ening to prevent this squeal or slippage will result in overloading the bearings or shaft. Belt speeds are normally limited to 5000 ft per min for E-section belts and 6500 ft per min for 8-V- section belts. Speeds in excess of these limits should not be used without consulting the belt manufacturer.

Gear drives require accurate alignment and rigid mounting. Pitch diameter and width should not be outside recommended gear manufacturer's limits. Check the factory for bearing thrust capacity before installing helical gears. In all cases, gear teeth must be centered with each other, correct shaft center distance must be attained, and gear faces must be parallel. Gear teeth must fully engage to a depth giving approximately 0.002 in. minimum backlash; avoid engagement so deep that gears will deflect or bind. Test for proper alignment by rotating motor shaft by hand, checking driven gear for backlash through one complete revolution. Test backlash and face parallelism again after tightening mounting bolts.

Mechanical alignment

With few exceptions, a flexible coupling is used to connect the motor to the load. This type of coupling is designed to tolerate some misalignment; however, this misalignment can cause vibration and/or stress on the motor bearings. Consequently, the shafts in all coupled applications should be lined up with the same high degree of accuracy, regardless of the type of coupling or type of bearing used. There are several important steps to follow in attaining correct alignment of direct-connected drives.

The foundation for the motor and driven load should provide a permament fixed relationship of the motor with respect to the driven load. The foundation should provide a solid anchor that will maintain this fixed relationship after alignment is completed.

Position the motor on its foundation to obtain the correct spacing between the motor shaft and the driven shaft. This distance is specified by the coupling manufacturers and its usually in the range of ⅛ to ⅜ in.

This positioning, in the case of sleeve-bearing motors, should limit the axial movement of the coupling to keep the motor bearing floating off of the thrust shoulders. These bearings will not take continuous thrust. When positioning the shaft of motors with end play, the shaft should be placed at the midpoint of the end play. (Ignore the magnetic center indication.)

Adjust the position of the motor by the use of jackscrews, shimming, etc., until the angular and parallel misalignment between the two shafts is within the recommended limits as measured with a dial indicator with the motor bolted down. When adjusting the position of the motor, care should be taken to assure that each foot of the motor is shimmed, before the motor is bolted down, so that no more than a .002-in. feeler gage can be inserted in the shim pack.

Angular misalignment is the amount by which the *faces* of the two coupling halves are out of parallel. It may be determined by mounting a dial indicator on one coupling half with the indicator probe on the face of the other half, then rotating both shafts together through 360° to determine any variation in reading.

It is important, during this check, to keep the shaft of a motor with end-play against its thrust shoulder and the shaft of the driven load with end-play against *its* thrust shoulder to prevent false readings due to shaft movements

---

Checkpoints for good motor support

1. No heavy welding should be done on machined bases after machining. If you see anything other than light tack welds on such a base, expect that some warpage has occurred.

2. Base welding should be continuous. Intermittent welds allow twisting and distortion in many cases.

3. Box sections or tubes are much stiffer than channels or beams alone. But make sure box sections are not "split" or left open along one side; these lose much of their resistance to twisting.

4. A one-piece base under a drive is always better than a sectional base.

5. Deep bases are better than shallow ones.

6. Watch especially for twist and distortion caused by overhung loads in bases above floor or ground level.

7. Adjustable braces are usually out of adjustment.

8. Don't take it for granted that the floor itself is solid. A solid-drive base on a resonant floor can be just as shaky as a weak base on a rigid floor.

9. Vibration during operation is a trouble indicator, and very often the trouble is in the drive base. Look at base or alignment first, the motor itself last.

10. Make sure *all* hold-down bolts are in place, and are *tight*.

11. Check to see that all separable base joints are doweled—motor-to-base, base-to-foundation beneath.

12. Watch for unshimmed gaps; all shims should be properly placed and in good condition.

13. Watch for stiffeners near attachment bolts. These greatly increase motor-base rigidity.

14. Look for bracing directly under the motor where it will do the most good; soleplates should be supported at least every 18 in.

---

in the axial direction.

Parallel misalignment is the offset between the centerlines of the two shafts. It may be determined by mounting a dial indicator on one coupling half with the indicator probe bearing radially on the other coupling half, then rotating both shafts together through 360°.

It is essential that the motor and load be correctly aligned under actual operating temperatures and conditions. Machines correctly aligned at room temperature may become badly misaligned, due to deformation or different thermal growth as they increase in

temperature. The alignment must be checked and corrected if necessary after the motor and driven machine have reached their maximum temperature under load.

It is recommended that "floating shaft couplings" or "spacer couplings" be used on motors where the coupling alignment cannot be accurately checked and/or maintained. Misalignments of several thousandths of an inch will result when there are relatively small changes in the temperature differences in larger motors and the equipment driven.

After alignment with the load, bolt

the motor in place with maximum-size bolts. It is advisable to provide some variation in the location of the foundation bolts. This can be done by locating the bolts in steel pipe embedded in the foundation. It is recommended that a competent engineer, familiar with motor foundation designs, be called upon to design and supervise foundations and support assemblies for large motors.

Next, the equipment should be given a test run to verify that the lineup gives satisfactory performance. Once satisfactory performance has been verified, the machines should be doweled to their bedplates. Recommended doweling is two dowels per machine, one in each of the diagonally opposite feet, with the size of the dowels approximately ½ the diameter of the hold-down bolts.

Machines correctly aligned when they are first installed may subsequently become misaligned due to wear, vibration, shifting of the base, settling of the foundations, thermal expansion and contraction, or corrosion. Therefore, it is advisable to recheck the alignment periodically to correct for any changes.

Electrical connections

Article 430 of the NEC and NEMA standards provide specific electrical and mechanical installation requirements and recommendations covering motors and motor controls.

Be sure that the voltage supply to the motor meets the motor requirements. Characteristics of the supply should correlate to the motor nameplate values as follows:

1. Voltage—within 10% above or below the value stamped on the nameplate.

2. Frequency—within 5% above or below the value stamped on the nameplate.

3. Voltage and frequency together—within 10% (providing frequency above is less than 5%) above or below values stamped on the nameplate.

After it has been ascertained that supply-voltage requirements are correct, the motor-terminal connections should be made. Stator winding connections should be made as shown on the nameplate connection diagram or in accordance with the wiring diagram attached to the inside of the conduit box cover. Terminal connection problems are usually caused by the branch-circuit conductors being a size that is

different from that of the motor leads. Branch-circuit wire size is normally determined in accordance with the NEC based on the motor full-load current, increased where required to limit voltage drop. The motor leads, on the other hand, are permitted a higher current-carrying capacity for a given AWG size than equivalent conductors used in branch-circuit wiring because they are exposed to circulating air within the motor.

When connecting the motor terminals to the line leads, use connector lugs sized to the conductor. Lugs so chosen, however, may not connect to each other securely. There may be difficulties in matching the two sets of lugs so that the full surface area of the small terminal can contact that of the larger. Washers should never be used between the motor lugs and the branch-circuit lugs in an attempt to promote better contract.

A higher degree of reliability is possible if connections are tightened according to torque specifications. Proper terminations can be made up with a torque wrench. Recommended torques are as follows:

Bolt diameter (in.)	Tightening torque (lb-ft)
¼	8
⁵⁄₁₆	14
³⁄₈	20
½	40

For wound-rotor motors, make connections to the rotor circuit in accordance with the wiring diagram furnished with the control apparatus. Also, conductors from the secondary control to the brushholder assembly must be installed and connected.

Brushes must make good contact with slip rings along the whole face of the brush. If needed, grind brushes in with fine sandpaper. Do not use lubricants. Service ports are provided for inspection of slip rings and brushes. Spring pressure should be set at about 1 lb on 215-286T frames and 4 lbs on 324-449T frames.

Be sure the motor terminal box is of sufficient size to permit good, reliable connections. Over the years it has been revealed that the most frequent single complaint of those who work with motors is "the terminal box is too small." In the 250- to 1000-hp range, particularly, the type and size of incoming cable determines how much room is needed. The designer is seldom given that information. The NEC is of

no help because it offers so many options, such as the choice of copper or aluminum cable. In working more closely with the actual installation, the installer usually is in a better position to select the most appropriate box.

When mounting conditions permit, the motor terminal box may be turned so that entrance can be made upward, downward, or from either side. For oversize conduit boxes, such as those required for stress cones or surge-protection equipment, the mounting height of the motor may have to be increased for accessibility.

It is recommended that the motor leads be supported and protected. They should be clamped at or near the point where they enter the box. A neoprene-sheet "lead separator gasket" is helpful. This limits movement of cables under starting currents. Extra sleeving around leads in this area eliminates chafing or cutting of cable jackets. Holes in steel plates through which leads pass ought to be chamfered or at least deburred.

Big, heavy boxes need extra support from the motor frame or base. This is most likely with any box over 18 to 20 in. high, especially if it contains heavy accessories. When you start up large drives, watch the light reflected from flat terminal box surfaces. You may see a "shimmer," indicating box vibration of fairly high amplitude, even when motor frame vibration is quite low. Such shaking may eventually crack the box, break bolts, or damage the leads. When in doubt, brace it.

Identify motor auxiliary devices such as space heaters or temperature sensors. Connect these in proper circuits and insulate from motor power cables.

Startup

After installation is completed, but before the motor is put into regular service, make an initial start as follows.

1. Check that the motor, starting, and control-device connections agree with wiring diagrams.

2. Be sure that voltage, phase, and frequency of line circuit (power supply) agree with the motor nameplate.

3. Check motor service records and tags accompanying the motor. Be certain bearings have been properly lubricated and oil wells are filled.

4. Perform insulation-resistance tests. Before carrying out the test, the machine must be at a standstill, and all windings to be tested must be electri-

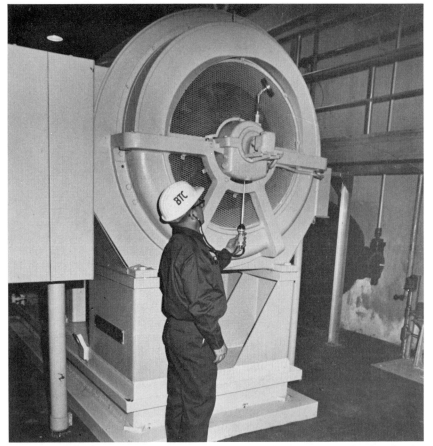

FINAL STAGES OF INSTALLATION includes numerous checks and tests. Bearing vibration is being checked with a special stethoscope designed for the purpose. If excessive noise or vibration is detected, complete vibration analysis will be performed on the motor.

cally connected to the frame and to ground for a time sufficient to remove all residual electrostatic charge. Failure to observe these precautions may result in injury to personnel. In accordance with established standards, the recommended minimum insulation resistance R_S for the stator winding is given by $1 + (V_S/1000)$ where R_S is in megohms at 40°C of the entire stator winding obtained by applying direct potential to the entire winding for one minute, and V_S is rated machine voltage. (See IEEE *Recommended Practice for Testing Insulation Resistance of Rotating Machines*, Publication No. 43, for more complete information.) If the insulation resistance is lower than this value, it is advisable to check for and eliminate moisture in accordance with accepted procedures.

5. Be sure that the rotor turns freely and does not rub when disconnected from the load. Any foreign matter in the air gap should be removed.

6. If the drive is disconnected, start the motor at no load long enough to be certain that no unusual condition

exists. Listen and feel for excessive noise, vibration, clicking, or pounding. If present, stop the motor immediately. Investigate the cause and correct before putting the motor in service.

7. Check direction or rotation. If incorrect, interchange any two line leads to reverse rotation on 3-phase motors. Fans on fan-cooled motors that have directional rotation nameplates must be reversed on the shaft if rotation is changed.

8. If the drive cannot be disconnected, interrupt the starting cycle after the motor has accelerated to low speed. Carefully check for any unusual conditions as the motor coasts to a stop. Repeat several times if necessary; however, repeated starts can overheat the motor. Refer to NEMA MG-1-20.42 and MG-1-20.43 or consult the factory for specific starting limitations to prevent damage to motor.

9. When these checks are satisfactory, operate the motor at lowest load possible and look for any unusual condition. Increase load slowly to maximum, checking unit for satisfactory

operation. Then run at normal speed watching carefully for any abnormalities.

10. Record all pertinent data, such as FL amps, volts, starting current (if an ammeter with the peak-current lock feature is available), insulation resistance, noise level, and temperature, for future reference.

Hazardous locations

Motors suitable for use in hazardous locations are tested and listed by UL in its *Hazardous Location Equipment Directory*. These motors, which are supplied with a UL Class I, Group C or D, or Class II, Group E, F, or G label, have been designed and manufactured in accordance with standards established by UL for explosionproof (Class I) and dust-ignitionproof (Class II) machines. Parts are machined to very close tolerances, including conduit box and/or collector-ring access covers. Extreme care must be taken during disassembly and reassembly, since any nicks or burrs may destroy the explosionproof or dust-ignitionproof features of the machine. If these features are altered in any way the machine will no longer comply with the provisions of the UL inspection and label service manual and will no longer be properly classified as a UL-labeled motor. The label should therefore be removed and the motor considered unsafe for use in hazardous locations. Always consult the manufacturer or a listed apparatus repair firm to assure safe and proper assembly. Motors for use in hazardous locations must be marked with an identification number that indicates the operating temperature range.

To date, UL lists no motors for Groups A and B; hence where such conditions are encountered, motors must be located outside the hazardous area. Motors suitable for other Class I locations are designated as explosionproof.

Likewise, motors for use in Class III locations are not listed. Class III locations are those where there is presence of ignitible fibers or flyings, such as in textile mills and woodworking plants. However, totally enclosed nonventilated motors and the so-called lint-free or self-cleaning textile squirrel-cage motors are commonly used. The latter may be acceptable to the local inspecting authority if only moderate amounts of flyings are likely to accumulate on or near the motor, which must be accessible for cleaning and maintenance.

Nameplate data—for better motor installation and maintenance

By E.J. FELDMAN, Senior Application Engineer,
Siemens-Allis, Inc., Norwood, OH

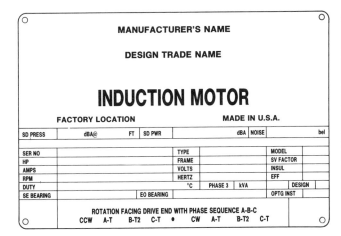

TWO TYPICAL NAMEPLATES, each for a different type of induction motor, show how data is presented. The nameplate at right indicates that the motor is a dual-speed type and includes terminal wiring diagrams. The left nameplate is for a "quiet" motor and provides information concerning the motor's noise level.

MOTOR NAMEPLATES provide a great deal of useful design and performance data. This information is particularly valuable to installers and to facility electrical personnel who maintain and replace existing motors. During installation, maintenance or replacement, the information on the nameplate is vital to fast and proper performance of the job at hand.

NEMA Publication MG1, Section 10.38, states that the following data should be stamped on every motor nameplate: Manufacturer, Type, Frame, Horsepower, Time Rating, Ambient Temperature, RPM, Frequency, Phases, Rated Load Amps, Voltage, Locked Rotor Code Letter, Design Letter, Service Factor and Insulation Class. In addition, the motor manufacturer may include such data as his name and manufacturing plant location and individual identification.

Most data given on the nameplate refers to electrical characteristics of the motor; therefore it is important for the installer/maintainer to be a qualified plant electrician/engineer or an industrial-oriented electrical contrac-

tor. For a better understanding of this data, two typical nameplates are illustrated here. The following discussion provides an explanation of the imprinted notations.

Serial Number: This is the individual, unique number assigned to this specific motor or design for identification should it be necessary to communicate with the manufacturer for any reason.

Type: The combination of letters and/or numbers chosen by the manufacturer identifies the type of enclosure and any significant modification thereof. It is necessary to have each manufacturer's coding system to understand the meaning.

Model Number: Additional identification data for manufacturer.

Horsepower: Rated horsepower is the horsepower the motor was designed to deliver at its shaft with rated frequency and voltage applied to the terminals with a service factor of 1.0.

Frame: Frame size designation identifies the dimensions of the motor. If it is a NEMA frame, it identifies the mounting dimensions (given in MG1) so that the manufacturer's dimen-

sioned drawings are not required.

Service Factor: The most common service factors are 1.0 to 1.15. If it is 1.0, it means that the motor should not be required to deliver greater than rated horsepower if injury to the insulation system is to be avoided. For 1.15 (actually, any value above 1.0), the motor may be run at a horsepower equal to rated horsepower times this service factor without serious injury to the insulation system. However, it should be pointed out that continuous operation within the service factor range will result in shortening the life expectancy of the insulation system.

Amps: The current drawn by the motor at rated voltage and frequency with full rated horsepower delivered to the load will be stamped in this space.

Volts: The design value of voltage is

stamped in this space. This should be the value as measured at the terminals of the motor rather than the voltage of the supply line. Standard voltages are given in MG1-10.30.

Class of Insulation: The class of insulating materials used in winding the stator is given here. These are systems of materials that have been extensively tested for long life exposed to preset temperatures. Class B system maximum operating temperature is 130° C; Class F system is 155° C; and Class H is 180° C.

RPM: This is the speed of the output shaft when delivering rated horsepower to the driven device with rated voltage and frequency applied to the terminals of the motor.

Hertz: This is the frequency of the supply system for which the motor is designed. Motors can operate at other frequencies, but performance will be altered.

Duty: Either "Intermittent" or "Continuous" is stamped in this space. If "Continuous," it means that the motor can be run 24 hrs/day, 365 days/year for many years. If "Intermittent," a time interval will be shown. This means the motor can operate at full load for the time interval given. After that time, the motor must be de-energized and allowed to cool before restarting.

Temperature Ambient: This specifies maximum ambient temperature, in °C, at which the motor can safely deliver rated horsepower. If ambient is higher than the stamped value, the motor output must be derated to prevent injury to the insulation system.

Phase: This entry indicates the number of phases for which the motor is designed. It must match the supply system.

kVA Code: The starting inrush current can be determined from this space. It is specified as a code letter indicating a range of kVA/hp. The range for each letter is specified in NEMA MG1-10.36. A common value is Code G, which covers a range of 5.6 up to, but not including, 6.3 kVA/hp. It is necessary to check the starting equipment for design compatibility as well as the local power company acceptance of this load on their system.

Design: When applicable, the NEMA design letter is stamped in this space. This letter specifies minimum torque values at locked rotor, pull-up and break-down speeds, and maximum inrush current and maximum rated

load slip value. These values are specified in NEMA MG1, Sections 1.16 and 1.17.

Bearings: For motors supplied with antifriction bearings, the bearings are identified by stamping the appropriate sequence of numbers and letter per AFBMA (Anti-Friction Bearing Manufacturers Assn.) Standards. Thus, the bearings can be replaced with the original design should replacement be necessary, since the AFBMA number includes the looseness of the bearing fit, the type of retainer, the degree of protection (shielded, sealed, open, etc., as well as the dimensions. Both the shaft end (SE) and end opposite (EO) the shaft bearings are designated.

Phase Sequence: The inclusion of phase sequence markings on the nameplate enables the installer to connect the motor for the designed rotation on the first attempt assuming the phase sequence of the supply line is known. If the supply sequence is A-B-C, connect leads as given on plate. If sequence is A-C-B, connect them opposite to that shown on the plate.

External connections are usually not shown on the nameplates of single-speed, 3-lead motors. However, for motors with more than three leads, the external connections are shown on the nameplate. Note that the nameplate for the dual-speed motor shows connections for low speed and for high speed. To run on low speed, line 1 must be connected to the motor lead T-1, line 2 to T-2, and line 3 to T-3, with motor leads T-4, T-5 and T-6 remaining open. For operation at high speed, connect line 1 to T-6, line 2 to T-4, and line 3 to T-5, with T-1, T-2, and T-3 shorted together.

Efficiency: This space will be marked with the NEMA nominal efficiency of the motor, which is derived from Table 12-4 of MG 1-12.53b. This efficiency data is applicable to "standard" motors as well as to "premium efficiency" motors. Those special-design, "energy-efficient" motors will be marked as such on the nameplate.

Low Noise: Some motors are designed for low noise emission. Therefore, the noise level is given on the nameplate expressed both as *sound power* and *sound pressure.* Both are measured in "dbA" units. This means the sound output from the motor has been fed through an "A" weighting network of filters which approximates the sound as perceived by the human ear. Since *sound pressure* has directiv-

ity, a distance must be given at which the measurement was taken. The accepted standard is 3 ft (or, more exact, 1 meter). These noise designations mean that the value stamped on the nameplate is the average level of noise measured on a hemisphere whose radius is equal to the longest dimensions of the motor (usually parallel to the shaft) plus 3 ft after passing through the "A"-weighted filter network.

NAMEPLATE (arrow) mounted on enclosure of 500-hp, 4000-V induction motor provides extensive information concerning installation and maintenance of the machine.

Sound power is a measure of the total energy emitted by the source. It is preferred for acoustical analysis since it can be combined mathematically with other sound sources to determine overall levels.

Noise emission is another way of stating the sound power in units outlined in Acoustical Society of America Standard 5-1976, which also references ANSI Spec. 51.23-1976.

Grounding motors effectively

By R. L. NAILEN, P.E.

MOST ELECTRICAL PERSONNEL are well aware of the value of a good equipment grounding circuit. Grounding building power distribution systems is comparatively easy; good guidance is available in existing codes and standards. However, effective grounding of rotating machinery often is not achieved because of a lack of such information or accepted practices.

If good equipment grounding of a motor or other rotating machine is to be provided, the objectives of such grounding must be fully understood. The potential between noncurrent-carrying parts of the equipment and between those parts and earth must be limited to a safe value under all conditions of both normal and abnormal operation, and a low-impedance return path must be provided for ground-fault current. A high impedance would permit dangerous voltages during a fault and could result in improper operation of protective devices. Also, high impedance at joints and connections may cause arcing and heating of sufficient magnitude to ignite nearby combustible materials or explosive gases.

Achieving these objectives involves the installation of a grounding conductor or path and a proper connection at the motor. The required size of a grounding conductor may be determined from Sec. 250-95 of the NEC, based on the rating or setting of the automatic overcurrent device in the circuit ahead of the equipment. The location and method of attachment of this grounding conductor to the motor is important. A terminal mounted with two bolts is more secure than one with one bolt, since rotation is prevented. However, it will take up more space, which could be a determining factor—particularly on smaller motors.

A split-bolt connector, threaded into the motor structure, could be used for the ground connection. On the other hand, a reliable connection may require a brazed-on or welded-on grounding pad or plate to provide a raised surface and thickness for bolt-type connections on the motor frame exterior or within

an enclosure or terminal box. There could be occasions when two (or more) such grounding pads are needed to provide for flexibility in routing grounding conductors from apparatus to selected grounding locations. Connection should be made to the largest mass of metal that is closest to the motor stator core assembly.

Very little guidance concerning motor grounding is available from industry standards. The most-effective grounding specifications, techniques and components come from operating practice and experience of user and consultants, tempered by good engineering judgment. The NEC, in Sec. 250-75, has this to say:

> Metal raceways, cable armor, cable sheath, enclosures, frames, fittings, and other metal noncurrent-carrying parts that are to serve as grounding conductors shall be effectively bonded where necessary to assure electrical continuity and the capacity to conduct safely any fault current likely . . . Any nonconductive paint, enamel, or similar coating shall be removed at threads, contact points, and contact surfaces . . .

Assembly bolts are normally steel, holding together steel parts that are invariably prime-painted (if not completely finished) prior to assembly. Generally, these components are thought of as structural components, not as current-carrying parts. However, it is entirely possible that they may be called upon to carry fault currents. If the ground path is to be effective, bolted joints must have low resistance. Each joint should have its steel surfaces ground smooth and clean to obtain good electrical contact. Serrated washers used with the terminal box mounting bolts may aid in making up good joints.

NEMA Standard MG 2-1977, entitled "Safety Standard for Construction and Guide for Selection, Installa-

HEAVY COPPER BUS (arrow 1) attached to one-piece frame of this 900-hp, 400-V pump motor assures a good "frame ground," But how well are the cable shields grounded within the terminal box at left? That depends on the bolted joints attaching that box (arrow 2) to the frame.

tion and Use of Electric Motors and Generators," says, in Section MG 2.09,

> . . . all exposed noncurrent-carrying metal parts which are likely to become energized under abnormal conditions shall make metal-to-metal contact or otherwise be electrically connected or bonded together to provide a common ground connection.

> When a motor or generator is provided with a grounding terminal, this terminal shall be on a part of the machine not normally disassembled during operation or servicing.

DICK NAILEN, a Senior Member of the IEEE, received a B.E.E. degree with honors from the University of Santa Clara, CA. Since 1953, he has been employed in electrical and mechanical design of rotating apparatus up to 19,000 kVA by major motor and generator manufacturers. He is a registered P.E. in Wisconsin.

Most modern motors have few parts "not normally disassembled" during servicing. Where then, should the equipment grounding connection be made? In most instances, the largest or heaviest portion of metal could be the motor end bell, the stator core assembly, or occasionally, the frame base. Theoretically, a grounding connection made into the stator core assembly would be effective. However, this assembly is made up of the frame and core laminations. Threading a bolt into these hard silicon-steel laminations would result in a poor joint. As the stator heats and cools throughout the life of the motor, such a joint could only become less effective. On small

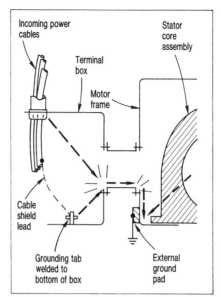

GROUNDING PATH on this high-voltage motor extends from conduit and cable shield through terminal box, throat, and motor frame assembly to an external solid ground. Arrangement shown has disadvantage of possible high-resistance joints at box/throat and throat/motor-frame locations. Arrows show possible path of ground fault or leakage current.

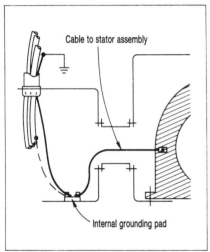

ONE EFFECTIVE grounding method is to use a single ground pad in the terminal box for solid grounding of all motor parts. A cable between the pad and the stator core assembly bypasses all the bolted joints; the grounding conductor in the conduit provides a low-resistance path to the closest ground point. This cable is bolted to the largest steel component in electrical connection with the core laminations.

machines, there is seldom room to make such a connection.

Grounding connections made to terminal boxes, particularly those provided with large and high-voltage motors, can be a source of problems. Boxes are of various sizes and are sometimes throat-connected to the stator. Alterations, test procedures, and maintenance work all may affect the grounding integrity. The result may be severe fault damage, costly downtime, or danger to operating personnel. Some users

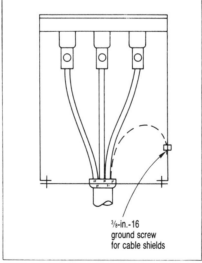

ANSI STANDARD C50.41 covering power-plant drive motors provides guidance for terminal-box grounding of shielded cables as shown here. Additional details that need to be provided include whether the ground screw is male of female and if it is intended to provide for connections both inside and outside of the terminal-box wall.

install their own boxes—perhaps the original box is found to be too small. Whatever the reason, it gets reworked or replaced. When that is done, the interconnecting joints, jumpers, etc., may not be replaced properly, or there may be more bolted joints than in the original assembly. As a result, terminal-box grounding paths may become obstructed, or high-resistance joints can be created. This is particularly true if motors are installed in wet, dirty, or corrosive surroundings. Too often, the best joint between steel parts may corrode quickly enough to interpose high resistance in the ground path.

Line connections that carry current continuously may loosen or corrode,

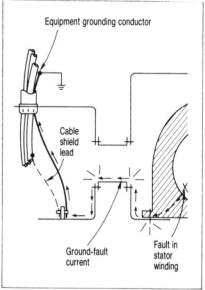

THIS CONNECTION relies entirely on the grounding conductor or conduit ground connection. With this arrangement, a ground fault in the stator winding is then subject to the possibly high-resistance path through the bolted joints at the stator/throat/terminal box.

and the damaging results (usually overheating or "blowout") will soon become apparent. But grounding connections normally carry no current, and the quality of the connection has no such self-checking property.

Many of these problems can be minimized by using grounding conductors extending from the closest good ground connection point via the motor supply conduit, with bonding jumpers between motor components. This may be accomplished in numerous ways, as shown in the accompanying photos and diagrams.

Watch out for resonance when installing pump-motor drives

By RICHARD NAILEN, P.E.

INSTALLATION of pump-motor drives often presents unusual and sometimes difficult mounting challenges. One such problem that can occur is mechanical resonance, sometimes called "tuning." Resonance appears when the natural frequency of the vibrations of the drive-system structure or foundation is the same or very close to the drive operating frequency, such as rotational speed. It is essential that the cause of this undesirable "tuning" be located and eliminated before vibration builds to an excess, causing damage that could conceivably result in total destruction of the drive.

A foundation may be "high-tuned" (natural frequency above running speed) or "low-tuned." But if it naturally vibrates right at running speed, trouble is inevitable. Would that be the motor's fault? Not at all. However, because shutting off the motor makes the trouble disappear, engineers tend to lay the blame there immediately—and wrongly.

Let's consider some examples of the problem. First, look at a 300-hp pump drive in a particular paper mill. Severe vibration was causing frequent motor-bearing failures. Plant engineers decided that the application was "too demanding" for a standard motor, so they issued specs for a "special design" to cope with the condition. Only later did they study the drive with a vibration analyzer. This showed the motor was oscillating at one end at a natural frequency of 3600 Hz/min—the same as the drive speed. Reason: the base plate was inadequately grouted-in at one end. Once this was corrected, the trouble disappeared.

As a second example: six 150-hp pump motors were mounted on steel beams 5 ft above the plant floor. Serious noise and vibration plagued the installation. Rebalancing and replacing the motors produced no improvement. Finally, a test of the foundation structure showed that the beams were

"tuned" to resonate at drive rpm.

Some engineers believe resonance can be eliminated as a drive problem by insisting on near-perfect balance of all rotating parts, so that if any vibration does appear later it must indicate a loss of balance that can be corrected. But such perfection is not attainable.

Structural resonance can be dealt with in only two ways. One solution to the problem is to "de-tune" the base using gussets, stiffeners, etc., to raise the natural frequency, or to cut away material or introduce spring washers in the motor mounting to lower that frequency.

A second solution can be to increase damping of vibrations by raising the mass of the drive. The vibrating structure represents work being done, because energy is required to move the parts. If more mass or inertia can be put into that vibrating structure, enough of the energy may be absorbed to limit the motion to harmless values.

Here's an example of the addition of mass to the vibrating system—an effective cure—but certainly one to be avoided by proper initial design of the base structure. Several 3000-hp 1200-rpm pump motors were installed on vertical-pump heads after passing all factory tests. At startup, they vibrated as much as 17 mils. Lack of pump-head stiffness was the cause. To change the assembled mass, shifting its natural vibration frequency as well as increasing damping, operators stacked 3000 lbs of sandbags on top of each motor. Vibration then dropped to only one mil. Until this was done, it was difficult to believe that the base stiffness was the source of the problem.

Vertical pump motors

Such problems occur with a particular frequency on vertical-pump drives. Because the motor is attached to its base only at one end, it is free to shake at the other end, unlike a horizontal motor with attachment points at both

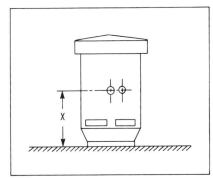

FIG. 1. Data needed to check natural frequency of assembled motor and vertical pump combination include location of motor center of gravity, X, and static deflection of that center that would occur due to motor weight if motor were mounted from its flange in a horizontal position. These data determine natural resonance or "reed frequency" of the motor per NEMA standards.

ends. But some pump drive specifiers have only recently begun to take into account the effect of how adding the motor alters the natural vibration frequency of the assembly. To specify this equipment properly, request from the motor manufacturer the following data, as outlined in NEMA standards:

1. Motor weight.

2. Center of gravity location. This is the distance from the motor mounting flange to the center of gravity of the motor.

3. Motor static deflection. This is the distance the center of gravity would be displaced downward from its original position if the motor were horizontally mounted. This value assumes that the motor uses its normal mounting and fastening means but that the foundation to which it is fastened does not deflect. See Fig. 1.

Another reason why vertical-pump heads are a frequent source of drive vibration is the type and number of cutouts in the head (Fig. 2). Cutouts are often provided in opposite sides for coupling access. These render the structure much less rigid in one plane than it is 90° away, giving the assembly two natural vibration frequencies in the two directions, one much lower than the other.

For example, the top-end vibration of one 800-hp, 1200-rpm vertical-pump motor was 30 mils in one direction and 13 mils 90° away. With the motor off the pump head, sitting on the floor,

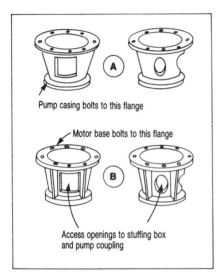

FIG. 2 Two types of vertical motor pump heads are shown. Those at (A), used for all sizes of motors, are basically a thin cone, weakened by the access opening cutouts. Those at (B) are stiffened by gussets or beam sections to greatly reduce the chance of serious vibration in the final assembly.

vibration did not exceed 2 mils in any direction. It was not feasible to stiffen the pump head. Instead, spring washers were used on the hold-down bolts to give the motor a "soft" mounting on the pump. This lowered the resonant frequency below operating speed frequency, resulting in assembled motor vibration of 6 mils maximum.

Pump heads tend to be of thin steel, with little reinforcement, even for large machines, apparently because designers feel the cylindrical shape will inherently be stiff enough. But this often is not the case. When you see pump heads such as shown in Fig. 2a, watch out. The use of stiffeners as shown in Fig. 2b is a great improvement.

Even with a stiff support, vibration

and even misalignment may be introduced through connected piping. Bolted and welded couplings of various types can transmit pressure surges or reaction forces into the drive. Consider the type of in-line pump shown in Fig. 3. Even slight vibration may set the assembly twanging like a bowstring. The pipe is strong enough to support pump weight but is lacking in stiffness.

More important, the attachment of heavy pipe—whether or not any fluid pulsation forces are present—can cause twisting or loosening of motor mountings. In one instance, identical pump motors side by side were found to have entirely different vibration levels. One was satisfactory, the other dangerously high. Loosening individual bolts in the pump discharge line coupling was enough to make the high vibration virtually disappear.

Fig. 4 shows a solution to this kind of problem. A heavy steel support was welded to the heavy discharge pipe to absorb the strong forces that occurred within the pipe. When any pump drive is being planned, piping forces should be carefully investigated.

Although small industrial motors are most often involved in foundation-vibration problems simply because there are so many of them installed under such widely varying circumstances, larger and more costly machines are not immune.

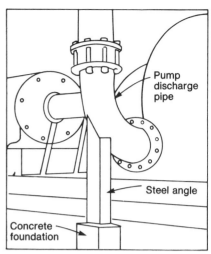

FIG. 4. Heavy steel angle welded to pump discharge pipe in foreground bypasses to a separate concrete foundation the piping forces that otherwise might be transmitted to the main drive foundation and machinery.

PUMP-DRIVE MOTORS are well supported on rigid steel base and concrete foundation built into floor. Heavy steel bracing at piping helps minimize forces from within the pipe.

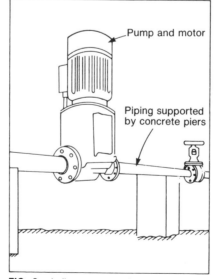

FIG. 3. In-line pump and motor are supported only by piping resting on concrete blocks. These blocks cannot be counted on to dampen vibrations originating in either piping or drive, which can be destructive if amplified by resonance.

Motor retrofits to save over $2-million

TWO UPSTATE New York industrial plants recently invested nearly $½-million in energy-efficient motors, replacing 286 "standard" motors in a program expected to save almost $2-million in the next decade.

One facility was the General Electric Silicone Products plant, Waterford, NY, which purchased 106 energy-efficient motors; the other was the GE Noryl plant in Selkirk, NY, which purchased 180 units.

Both plants are chemical/process-production plants that run continuously, or approximately 7000 hrs/yrs, which is a major factor in developing the cost justification for converting to the energy-efficient motors. At both plants, 80 to 90% of the power used is consumed by motors. Motors selected for retrofit ranged in size from 7½ to 250 hp and drive pumps, fans, compressors, and certain process machines.

Management at both facilities has been aware of the benefits of energy conservation since the mid '70s. Old outdated lighting systems have been replaced with new, efficient, high-pressure-sodium lighting. Boiler systems have been modernized, HVAC system operations has been made more efficient, and total energy management using microprocessor-based equipment is under consideration.

After seeing the dollar savings obtained from the initial electric energy-saving measures, management at both plants looked closely at their largest electric power consumers—their motors. At first, energy-efficient motors were specified as the required drive on newly purchased machines or plant equipment. Next, the energy-efficient motors were purchased as replacements for most failed motors. After actual field operation proved that significant savings were obtained, programs at both plants were launched to retrofit appropriate motors.

At the Waterford plant, the entire

30-HP ENERGY-EFFICIENT MOTOR is being checked by James Crandell at the GE Silicone Products Div., Waterford, NY. The motor is typical of 106 that replaced older, less-efficient motors. Electric energy savings are expected to be approximately $60,000/yr with a payback of slightly more than two years.

population of 1500 motors was surveyed, listing size, hours of operation, application, and priority of the process involved. Priority was important because plant electrical people, who are responsible for electrical maintenance at these continuous-process plants, called for maximum reliability of motors. In addition to running efficiently, these motors are highly reliable. Because they have better steel, improved characteristics in the stator and rotor, and extra copper in the windings, they operate at temperatures 20° to 30° C cooler than conventional

motors. As a result, insulation and bearing lubricant life is increased, meaning reduced maintenance and downtime, which is particularly expensive at these continuous-operation plants.

Of the 1500 motors at the Waterford plant, 106 met all the retrofit requirements. This included a payback period of about two years. At the local purchased power cost of about $.06/kWh, plant management expects to save nearly $60,000 on electric energy bills in the energy-efficient motor's first year of operation. This results in a payback period of slightly more than 2 years. Assuming a steady 10 to 15% rise in the electric energy rates, savings will total approximately $640,000 after 7 years.

At the Selkirk plant, over 3000 motors were audited. Approximately 300 were considered to meet initial criteria; however after careful analysis based on size, rating, application, and operating mode, 180 motors were selected to be replaced with new energy-efficient motors. Most chosen run at 7200 hrs/yr; others that run for less time were selected because of their advanced age and increased maintenance risk. These motors were the most obvious candidates for retrofit. They offer the quickest payback with the least disturbance of on-going factory operations. Later, additional motors will be changed out as they also fit the requirements of a retrofit program, which includes ease and speed of changeout, payback time, and noninterference with overall plant operation.

Purchased power at this location is also $.06/kWh, which will result in a savings of more than $100,000 in power costs the first year. With a projected 10.5% annual power cost increase, savings will total approximately $2 million after a decade. Payback is expected in just under three years.

Tackling the problem of reducing motor noise

A continuing program at E. I. du Pont has significantly reduced worker exposure to potentially harmful noise levels generated by totally enclosed, fan-cooled induction motors.

TEFC MOTORS, found in most industrial locations, are a source of noise that, if unabated, contributes to "noise pollution" of the workplace. **SOUND LEVEL** measuring instruments, such as that used here by author T. A. Dear, are designed to mimic the method by which the human ear responds to sound waves.

MOTORS are everywhere in industry, and so is the noise they make. Because the motor noise combines with the noise of the driven equipment, motors are often at the center of a cluster of problem-noise sources. For more than 20 years, the Du Pont Company has been concerned with reducing this noise.

In the mid-1950s, silencers were designed and fabricated for 75-hp, totally enclosed, fan-cooled (TEFC) motors located at one of its plant sites. A few years·later, commercial motor silencers became available and were applied to TEFC motors ranging from 50 to 125 hp. However, such silencers are bulky and expensive as retrofit hardware, so Du Pont engineers attacked the origins of motor noise. It was found that by improved motor cooling-fan design, significant and uni-

By TERRENCE A. DEAR, F.I.O.A.

form reductions in noise levels could be achieved. To understand why these design features do reduce motor noise, it's useful to review briefly how vibration of material and air in equipment becomes noise heard by the ear.

Sound and noise

Vibratory motion of a sounding body reaches the ear by means of "molecular" waves in the intervening medium; sound will not travel in a vacuum. The velocity of sound in air at 0°C is about 1090 ft/sec. A change in the temperature changes the speed of sound at the

T. A. DEAR has been a senior consultant for Du Pont's department for noise source engineering control and hearing conservation for the past 10 years. He holds more than 30 patents in noise control and is widely published and recognized for his work in acoustics.

rate of approximately 2 ft/sec/°C—increased temperature increases speed. Velocity of sound in denser media (such as water) is generally much higher than in air and is not so dependent upon temperature.

Because a sound wave is actually transmitted by the vibration of molecules of media in which it is traveling, it can eventually exert oscillatory pressure on the eardrum of a listener in range. As the molecules react in the direction of incidence of the wave, the pressure forces the eardrum inwards. Rarification of the medium (usually air) brought about by the reversing molecular motion causes the eardrum to respond in the opposite direction. Thus, the eardrum vibrates at the signal frequency (or frequencies) of the sound wave. The internal mechanism of the ear converts the vibration of the eardrum to electrical impulses in the brain and the sound is "heard."

Sound waves are often represented by sine waves (Fig. 1). Peaks represent compression of the air molecules; troughs indicate rarification as the vibrating source moves in the opposite direction from the initial sound wave. The wave length of the sound wave is the distance between two repeating points on the wave, and the amplitude is the magnitude of the sound pressure above or below the local barometric value. The rms (effective) value of amplitude is used in describing steady-state sound pressure. Frequency of the wave is found by dividing the velocity by the wave length.

The amplitude of a sound wave determines the loudness of sound and the energy in the wave. Frequency and pitch are analogous terms—the higher the frequency, the higher the pitch.

A musical tone is made up of regularly repeating or periodic vibrations. The sound of the plucked, vibrating string on a guitar, for example, is amplified by a properly tuned wooden case. But noise, defined as "unwanted sound," is generally associated with irregular vibrations. Sounds reaching the ear in most industrial locations do not consist of pure tones. They are a mix of sounds from sounding bodies and/or interfaces, reflected sound, transmitted sound, and sound from items that are in sympathetic vibration with other sources. For instance, the surface of a motor that is vibrating will cause noise. The cooling fan causes turbulence in the surrounding air and also generates noise. If the motor is

located near a poorly braced plywood wall, the sound waves will cause the plywood to vibrate at the same frequencies, and because it has a greater surface area than the motor, it can greatly amplify the sound level. Such sounds are clearly "noise."

Adverse effects of noise

Noise can annoy those exposed to it, and there is potential physical damage associated with it. Binaural deterioration of hearing that cannot be explained by the normal aging process or disease is often due to long-term exposure to excessive noise. It is not unusual for audiometric tests to show that some 25-year-old employees who spend hours at discos or riding motorcycles come on the job with 40-year-old ears.

In 1969, the federal government perceived noise exposure to be of sufficient importance to justify specific coverage under the Walsh-Healey Act. At that time, a noise exposure limit was imposed. Limits were placed on the number of hours an employee could work in a noisy location, the number of hours depending on the sound level (see Table 1). This was later reinforced by making these exposure levels part of OSHA regulations in 1971, thus extending coverage to all employees of a business engaged in or affecting interstate or foreign commerce.

Measuring sound

Reducing noise from motors, or from any other source, begins with measuring levels and identifying contributors. Sound pressure is measured in units of "micropascals." The minimum sound pressure that the young human ear can detect is about 20 micropascals at 1000 cycles/sec. But the listener's ear does not respond to an increase in sound pressure in a linear fashion. Twice the sound pressure is not detected as a sound that is twice as loud. The response is logarithmic. A decibel scale expresses the logarithm of the ratio of a measured quantity to a reference quantity. Thus, this scale has been adapted to the measurement of sound. The level of sound above the reference level—the threshold of hearing, 20 micropascals—is expressed in decibels (dB). Fig. 2 shows a typical decibel scale.

Sound-level measuring instruments generally consist of a microphone to pick up and convert the sound pressure to electrical impulses, signal processing

Fig.1. Sound wave as a sine wave

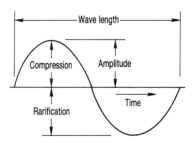

A vibrating body compresses and rarifies the molecules of air around it. Sound travels as each succeeding molecule is made to vibrate at the same frequency as the original body.

electronics, and a ballistic meter calibrated in decibels. They thus mimic, in most ways, the manner in which the human ear detects sound.

There is one additional complicating factor: the ear is not equally sensitive to sound at all frequencies. For this reason, even though the sound-pressure level of two different noises may be the same, one may be considered louder than the other if one is concentrated in the frequency region to which the ear is more sensitive. To allow for this, sound-level meters have several frequency-weighting networks that alter the sensitivity of the meter with respect to the frequency. The most commonly used one is called the "A-weighting" network. This network is specified by OSHA regulations for sound-level measurements, and measurements taken using this A scale are expressed in dB(A) or simply dBA.

There is a natural time delay in the ear's response to sound. Rapid fluctuations are thus "time-averaged." To approximate even more closely the response of the human ear to sound, two response times are built into the meters. There are "fast" and "slow" responses. The first has a time constant of $\frac{1}{8}$ sec and the other 1 sec. OSHA specifies the "slow" response for sound-level measurements.

Enclosures and damping

Early attempts at motor noise abatement were based upon a generally accepted rule-of-thumb setting the maximum allowable human noise ex-

Table 1. Permissible noise exposure (Walsh-Healey and OSHA)

Duration per day (hours)	Sound level dBA slow response
8	90
6	92
4	95
3	97
2	100
1½	107
1	105
½	110
¼ or less	115

When the daily noise exposure is composed of two or more periods at different levels, their combined effect should be considered. If the sums exceed unity, the mixed noise exposure should be considered to exceed the limit value. Exposure to impulse or impact noise should not exceed 140 dB peak sound pressure level.

Fig. 2. Relation between sound pressure in pascals and sound pressure in decibels referenced to 20 micropascals

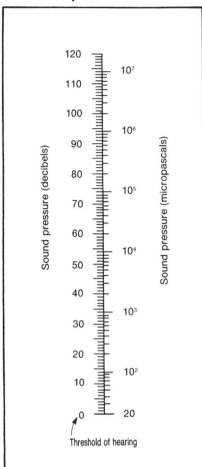

posure level at 97.5 dBA. To allow for noise contributions from other sources and for reflection, individual sources were permitted to generate noise levels no greater than 91.5 dBA. To meet this requirement, it was often necessary to provide sound treatment after the motor was installed, sometimes called retrofit. Typically, a 250-hp, 3600-rpm TEFC motor was provided with a covering structure (enclosure) made up of angle iron, 18-gauge sheet steel, and a 1-in fiberglass lining. Inlet sound-absorbing ducts were provided for cooling air and exhaust air. Another method of quieting units is the isolation and damping of vibration using material or hardware to prevent transmission of sound through floor slabs, etc.

Reducing noise at source

Du Pont's noise program focuses on control of noise at the source as a long-range approach to achieving lasting benefits. Origins of noise in motors are usually points where local pressure perturbations are caused by air turbulence and forced motion of component surfaces. By reducing the noise level at the source through proper design modifications, additional benefits can be gained: vibration will be reduced,

equipment life extended, and maintenance reduced.

Noise from TEFC motors has its primary origins along the surfaces of the motor cooling fan as it cuts through the air. As might be expected, the highest noise generation is along the tips of the blades, where the relative velocity and air turbulence are at a maximum. Attacking this origin of motor noise by engineering design showed that there were numerous modifications that could be effective—slower tip speed, improved inflow paths, improved motor insulations, and changes in fan blades and housing structures.

A review of one case history points out other approaches that can be taken in dealing with motor noise through design. Initial tests on two large induction motors showed sound levels in the high 90-dBA range. Because of other noise contributions expected in the

work area in which they were to be installed, they had been specified to be in the low 80-dBA range.

Thorough investigation pointed to three major noise contributors. One proved to be a force wave at twice the line frequency causing the stator to deform in a near-sinusoidal mode around the periphery of the core. This set up vibration-induced resonances in the rotor and stator cores that were efficiently coupled to the motor structure. In addition, components directly in contact with the surrounding air contributed to radiated noise by generating turbulence. A second high-frequency noise was due to air-gap flux harmonics caused by pole pairs differing by a relatively small integer.

Of primary importance in this instance was a third noise generated by harmonic waves produced by the interaction of rotor and stator slots. The actual analysis of these waves is quite complex, but it is sufficient to say that the interaction between them and the main air-gap waves produced forces with identifiable frequencies and number of nodes. Increasing the number of rotor bars was the solution in this case. Changing the number of rotor bars for all induction motors will not necessarily provide the same results; in fact, this approach could intensify the problem in some cases. Each case must be considered on an individual basis. In the example cited, redesign was accomplished by the manufacturer. The noise level of these large, high-horsepower motors dropped below 80 dBA with corresponding reductions in core vibration levels to 1/50 of the original accelerations.

Specifying low-noise motors

Specifying low-noise motors is becoming more common. A vendor can be expected to meet noise specifications as well as any other motor performance criterion.

Additive effects of noise levels in work areas create the need for a wide range of available motor sound power levels at the same horsepower rating. The specified sound level depends upon the fractional contribution permitted for each motor (or any other source) in a particular workplace. Table 2 indicates some typical noise levels of both standard and low-noise motors, which vary by manufacturer. Assuming that the standard values fall within the requirements of the workplace composition of noise sources, they can be

Table 2. Typical TEFC sound power levels (dBA)

Motor Rating (horsepower)	Standard		Low-noise type	
	3600 r/min	1800 r/min	3600 r/min	1800 r/min
1-2	88	74	78	67
3-5	91	79	80	70
7.5-10	94	84	84	76
15-20	98	89	84	82
25-30	100	92	84	83
40-50	103	97	84	80
60-75	105	100	83	83
100	106	102	82	84
125-150	107	104	83	81
200	—	—	85	85

specified directly. Otherwise, detailed noise specifications must be written.

The values given in Table 2 are sound *power* levels. Power levels have replaced pressure levels as a standard for motor noise specification to remove acoustical differences in test and installation environments. The dBA levels are based on a hearing threshold of 1 picowatt (10^{-12} watt) rather than 20 micropascals. Sound-pressure level correspondingly depends not only on the sound power level of the source but also on the distance from the source and on the acoustical characteristics of the space surrounding the source. This refinement in measurement of sound power level is necessary to ensure that all motors claiming an 84-dBA sound level have indeed been tested under similar conditions. The purchaser thus can compare apples and apples instead of apples and oranges.

When specifications for specially fabricated low-noise equipment are written, they should include a data chart similar to the one shown. Using noise specification data determined according to good acoustical practice, all applicable requirements available at the time of purchase must be filled in. This information should include specified sound levels and microphone locations where measured sound-power levels are not available.

Test verification of sound levels should be entered in the data chart side-by-side with the specified data. The report format should include such considerations as acoustical instrumentation used in the test, acoustical characteristics of the test facility, location of equipment within the test facility, description of equipment mounting and installation for test, description of interfacing equipment, test loading of equipment, and microphone locations relative to the unit, as appropriate, considering the differing approaches to sound power level and sound pressure level measurements and related data documentation. All data sheets should be properly filled out, dated and signed by persons conducting the tests and by the witnesses designated by the purchaser. Failure to achieve noise specifications should be considered to be as relevant as any other equipment-performance deficiency.

With quiet motors, industrial noise levels potentially harmful to workers can be reduced by source engineering controls. Through research, testing, and cooperation with vendors—and by specifying low noise levels for motors purchased—Du Pont is continuing its program to improve noise control in the workplace. This approach would seem well suited for noise reduction of other types of machinery as well, even though it takes a firm commitment and many years to accomplish.

DATA CHART

Sound pressure ☐   Level specifications (S)
Sound power ☐   and verification test measurements (M)
Load condition: full ☐ or_____ %   Flow _____   Pressure _____

Meter setting	Background noise	1S	1M	2S	2M	3S	3M	4S	4M	5S	5M	6S	6M
dBA													
Octave band center frequencies (Hz) 31.5													
63													
125													
250													
500													
1000													
2000													
4000													
8000													

Mounting methods to minimize motor maintenance

By RICHARD NAILEN, P. E.

A GREAT MANY motor troubles originate in the way the motor is first installed. In many instances, the foundation or baseplate structure is either incorrectly designed or improperly built—or both. The inevitable result is vibration, shaft misalignment, bearing damage, even breakage of the shaft or of the motor frame itself—sometimes bringing on catastrophic electrical failure as well.

If the motor is to be mounted on a concrete base, it is essential that this foundation be *rigid* to minimize vibration and misalignment problems during operation. The foundation should be made of solid concrete with its footings set deep enough to be resting on a firm sub-base.

In the event that the motor must be mounted on steel, all supports must be of adequate size and strength and braced to assure maximum rigidity (Fig. 1).

Whether the motor base is concrete or steel, it must level. If concrete, be sure it is not too high. A foundation can always be raised by use of shims; but reduction of height would be difficult because it would require removal of some of the concrete surface.

The requirement for a level base is critical. Usually, for a motor installation, there will be four points of mounting—one at each corner of the mounting base. Then there will be mounting requirements for the driven load. All mounting points must be on the exact same plane or the equipment will *not* be level. This is why a thick, rigid steel baseplate is preferred over an assembly of steel, or at least a steel baseplate should be used in conjunction with the steel assembly. For installation of large and heavy equipment, it is recommended that the services of a civil or mechanical engineer be employed to assure proper installation.

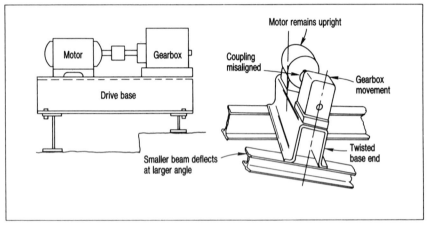

FIG. 1. Steel supports must be of adequate size and strength and braced against twisting, which can occur during operation of large high-torque motor drives.

Grouting and shimming

Proper grouting is of great importance to foundation stiffness and stability. Even the best steel baseplate structures aren't considered proper support unless they are grouted in place. Fig. 2 shows an exaggeration of what may otherwise happen. Correct selection, mixing, and placement of grout is important, and suppliers of such material should be consulted whenever this work is to be done. The accompanying checklist highlights some important proper procedures.

Correct shimming is another essential to good motor/foundation assembly. It is as important as base design itself; yet it may get little attention. One way to assure a proper shim installation is to physically remove and inspect the shims at every machine support just prior to final alignment. Obvious problems with shims include rust, improper cut, folds and wrinkles, burrs, hammer marks and dirt.

Remember why shims are used. They're not just to raise or lower the

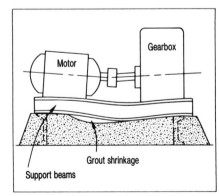

FIG. 2. Improper grouting can result in shrinkage that permits twisting of support beam, misalignment, and excessive vibration.

motor so shafts will line up. They also solve the basic geometry problem of insuring that all of the motor mounting pads or feet are in one flat plane. Should any one of them be higher or lower than the others, and the motor feet are bolted down tightly to all, either the base or the motor frame will have to be warped out of line.

One useful test for the source of troublesome drive vibration is to loosen motor hold-down bolts one at a time while the motor is running. If vibration noticeably decreases, the trouble is probably due to a distorted base or improper shimming.

For shims to do the job, they must form a solid, tight pack when the bolts are tightened. A "spongy" set of shims means a loose joint, which may or may not stay planar with other mounting points.

Another essential for a good foundation is stability (Fig. 3). Once properly built and installed, the base has to

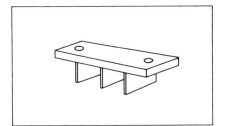

FIG. 3. Soleplate has "fins" welded to bottom to increase strength of its embedment into concrete foundation, resulting in more-rigid mounting.

remain that way. Excessive heat can be a problem. For example, too much welding heat is one way to throw the base out of kilter. In severe hot outdoor climates, there have even been reports of steel bases "twisting" when desert sun impinged on one side while the other side remained in shadow. If the base is common to both motor and driven load, this might not be serious because both drive elements will see the same movement. Machinery con-

nections such as heavy piping, however, must allow for this shift in position. If this isn't feasible, a sun shade may be needed to equalize base heating.

For many drives, alteration of bases in the field has meant trouble. In one instance, a steel bedplate had been used for years with a particular motor-pump combination. To meet new operating conditions, a different, smaller motor was installed. Its shaft height was matched to the existing pump by welding supporting "posts" at the four corners of the old bedplate. These were strong enough to hold motor weight but were too flexible to keep the motor from vibrating back and forth. Moreover, the welds weren't tight, so that one corner broke loose. Eventual result: bearing failure. At first, the problem was blamed on the motor; however, closer examination revealed the improper mounting methods that were used.

INSULATION STANDARDS AND APPLICATION OF MOTORS

Standards for motor insulation systems have been evolving in recent years to provide the most practical and effective method of specifying and applying motors. Latest insulation standards are based on specified test techniques that evaluate insulation based on simulated electrical, thermal, and mechanical stresses encountered by motors operating in actual service situations.

As a result, modern standards use a temperature-rise evaluation correlated to an ambient temperature of 40C. Values for various classes of insulating materials are determined by measurement and calculation of resistance of motor windings. These insulation classes for integral-horsepower motors are listed in the accompanying table.

Note that Class A insulation is not listed, since it is no longer used in integral-hp motors. Also note that the service factor of the motor and encapsulation affect the insulation temperature rating.

Motor type	Service factor	Temperature rise		
		Class B	Class F	Class H
Open & TEFC	1.00	80C	105C	125C
All motors	1.15 up	90C	115C	135C*
Encapsulated & TENV	1.00	85C	110C	—

TEFC—Totally enclosed, fan-cooled
TENV—Totally enclosed, nonventilated
*Rating is not standard but is acceptable to industry practice. Temperature rise is based on increase from 40C ambient temperature as calculated or measured change of winding resistance from startup to running at full load.

How ground fault systems protect motor loads

By GREGORY C. ECKART

MOTORS AND MOTOR CIRCUITS normally are well designed to provide adequate protection from short-circuit conditions and overloads. The damage or potential damage resulting from such faults are well known, and devices such as circuit breakers, fuses, and overload relays can be applied to obtain the needed protection. However, a third type of fault condition, a ground fault, can be just as destructive as an overload or short circuit and far more insidious.

Modern power distribution systems are typically rated 480/277 V and as such are susceptible to ground faults. These systems and motors connected to them can often be destroyed by ground faults if proper protection is not provided.

Today, many plants have ground-fault protection down to the feeders and large motors, but this may not be enough. Low-level arcing ground faults in unprotected downstream motors cause extensive motor replacement and repair costs. These faults cause severe deterioration of motor windings and laminations. Furthermore, low-level ground faults can escalate quickly and shut down motor control centers, which in certain cases may result in the entire manufacturing plant being shut down.

To provide the needed protection, new solid-state ground-fault protection systems have been designed and made available. As shown in the accompanying diagram and photos, a typical system consists of a current sensor (usu-ally a donut-type current transformer) and a multifunction solid-state relay. Usually, indication is provided on the relay; but remote monitoring is also available.

The ground-fault systems are used to protect single-phase and polyphase ac motors rated from 20 to 500 hp. They can also be used on larger voltage- and hp-rated motors but will probably require different current transformers than normally supplied with the ground-fault relays. In most cases, this will depend on the sizes and types of cables installed.

The devices are installed where the continuous use of a motor is critical to a manufacturing process or where finding a replacement for a burned-out motor is difficult. There is less need for ground-fault protection on fractional-hp or integral-hp motors below 20 hp. These motors are usually kept in stock or are easy to obtain.

Ground-fault protective devices are small, compact, and inexpensive; and they can be installed in existing motor control centers and combination starters. The ground-fault relays may be wired to operate a circuit-breaker

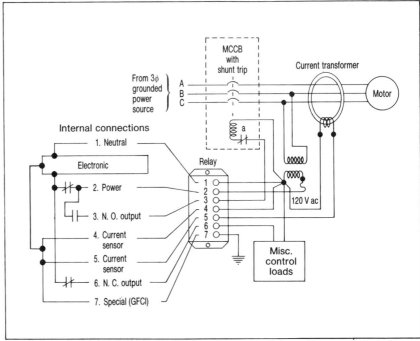

GROUND-FAULT PROTECTION system as installed on a motor circuit consists of a dc nut-type current transformer sensor that detects any unbalanced current in the supply conductors to the motor, an adjustable control relay with numerous functions as listed, a shunt-trip CB and auxiliary operating contact, and an integral GFCI protector.

GREGORY ECKART is a senior application engineer with General Electric Co.'s Distribution Equipment Div. in Plainville, CT.

shunt trip or to drop out a contactor. Normally closed contacts are rated 15 A continuous, 90 A make and 15 A break, and 120 V ac. Normally open contacts are 3 A continuous, 15 A inrush. They are suitable for use on systems capable of delivering up to 100,000 A rms phase-to-phase or phase-to-ground current.

A current transformer, which detects ground faults, serves as a signaling device and is smaller and more compact than conventional current sensors. Current sensors have a 6-step range to allow selecting the specific pickup setting that is correct for a particular application, from 50 mA to 30 A.

The relay, although very sensitive, has an inverse time-delay characteristic with an instantaneous unit for intermediate and high-level ground faults and has a very high resistance to nuisance tripping. Ground-fault-trip

ampere adjustment ranges available are: 0.05 to 0.3; 0.1 to 0.6; 0.5 to 3.0; 1.0 to 6.0; 2.0 to 12.0, and 5.0 to 30.0.

When selecting a trip setting, contractors, consultants and plant electrical people often choose "the lowest possible" setting. Others believe the trip setting is selected based on what the user feels is most comfortable. In any event, the proper setting is usually dependent upon three parameters: (1) the amount of damage the operator or owner is willing to permit, (2) whether there is more than one load on a circuit, and (3) whether a nonmotor load on the circuit, such as a heater, requires a higher than normally permissible setting.

To annunciate the occurrence of a fault, the system has an integral ground-fault target indicator. The protection system also has an integral test mechanism that enables the user to simulate a fault to make sure the sys-

tem is working.

In addition to protecting motors and the distribution system, the ground-fault relay can be wired to provide personnel protection (GFCI) on the 120-V circuit. Plant workers can be provided with ground-fault circuit interruptions at 5 mA, the level established by the NEC and Underwriters Laboratories for personnel protection.

In recommending circuit protection for motors against all three types of faults, the best approach would be a starter with overload relays, a magnetic-only circuit breaker, and a separate ground-fault relay. The overload relays protect both the cable and the motor. The CB provides short-circuit protection for the starter with overload relays, cable, and the motor; and the ground-fault relay protects motors on the 480V circuit and provides control-circuit ground-fault circuit interruption protection for personnel.

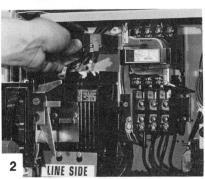

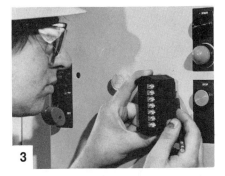

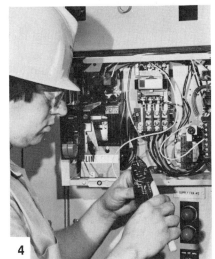

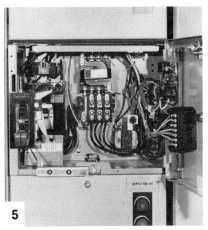

PHOTO SEQUENCE shows addition of motor ground-fault protection components to an existing motor controller.

1. Current transformer (left hand) and solid-state relay (right hand) are both small in size and easy to install.

2. Current sensor (transformer) is mounted above shunt-trip-equipped CB to permit 3-phase power source conductors to pass through sensor to starter.

3. Ground-fault relay position is outlined on door of motor enclo-

sure. Metal will be cut out to permit front of relay to be observed and operated with door of enclosure closed.

4. Flexible control conductors are connected between the current transformer and relay.

5. Completed installation shows relay on door and current transformer over circuit breaker.

6. Left button of ground-fault relay displays a red stripe when relay has tripped. Pressing button resets unit. Right button is for testing.

Cycling of motors for energy conservation

Revised NEMA Standards Publication No. MG-10 provides a guideline for the maximum permitted starts per hour and minimum off time between starts.

INCREASED COST of electricity since the 1978 energy crisis has popularized the practice of cycling on and off items such as HVAC and refrigeration-equipment motors. The energy savings gained by doing so must be balanced against the possible shortening of the life of the motor and its starting equipment. Until now there has been no generally accepted method for determining how many times per hour a motor can be started without significantly reducing its useful operating life. The only available guideline has been NEMA's Standard MG-1, Clause 12.50. It limited the number of starts to two from ambient or one start from rated-load operating temperature. However, this information is not applicable to repetitive start-run-stop-rest cycles used in energy-management programs. The new information contained in NEMA MG-10 *Energy Management Guide For Selection And Use Of Polyphase Motors* is intended specifically to provide the information needed to make these calculations for NEMA Design A and Design B motors.

Causes of cycling damage

Effective service life of motors is often determined not by the total number of hours that it has run, but by the number of starts it has been subjected to. There are two components that singly or in combination can cause a motor to fail due to cycling. The first is overtemperature, and the second is the high electromagnetic stresses imposed during the starting period.

The insulations of motor windings are particularly temperature-sensitive. A rule-of-thumb says that the life of the insulation is halved for each 8°-12°C increase of operating temperature above its rating. Because electric heating effect increases as the square of the

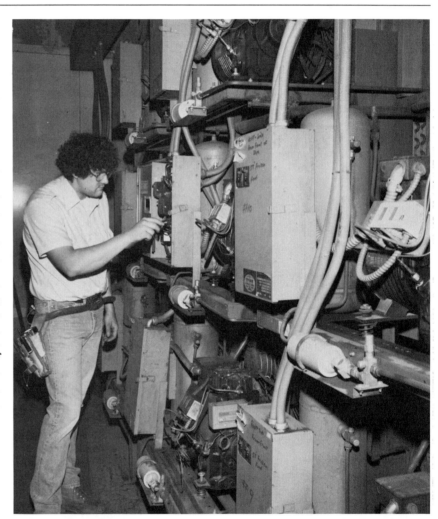

ELECTRIC MOTORS consume more electricity than any other equipment, so it's understandable that energy-management programs intended to reduce power bills often include provisions for periodically shutting off motors. In most cases this practice saves money on electricity bills but, depending on cycling frequency, it can also lead to shortened motor life. NEMA has produced a guideline that will help determine how many times per hour a motor can be started without significantly reducing its life.

current, the average six-times-normal FLA experienced during starting quickly raises temperatures within a motor. When a motor is started, the rate of the temperature rise in the stator and rotor winding is very high.

There is little heat transfer from the winding to the surrounding core. Additionally, shaft-driven ventilating fans are relatively ineffective until the motor is running at rated speed. As a result, several starts in succession will cause the temperature to exceed the thermal limit of the stator or rotor winding.

Insulation breakdown as a function of temperature is a chemical process that is accelerated by increased temperatures. However, heat has several other consequences. The resistance of the insulation *decreases* as temperature rises. Thus the windings are more prone to shorting. Also, the rotors of motors can be damaged. Because the rotor winding is part of the heat sink, its bars can melt or soften and be distorted by centrifugal and electromagnetic forces. Large motors driving equipment that requires a long time to accelerate to operating speed are more likely to experience rotor damage, but smaller motors are also vulnerable to this type of thermally induced damage.

Starting current also exerts strong repelling forces on stator-winding end coils. Even slight movement of the coils can, over a period of time, cause insulation failure or embrittlement of the coils themselves. This is further complicated by the fact that some insulations soften with elevated temperature and increase the ability of coils to move in response to the forces. Vibration, especially during long starting periods, adds to the possibility of coil mechanical failure.

How driven loads affect motors

Heating effect depends upon the amount of current required to start a motor. The longer it takes to accelerate a load up to operating speed, the more total heat will be released. Total starting time depends chiefly on the nature of the load that must be brought up to speed. The time to reach operating speed can be calculated from

$$t = Wk^2 \times (rpm/307.5) \times T_a$$

where Wk^2 is the inertia of the motor plus the driven equipment and T_a is the torque available for overcoming inertia (motor torque minus required load torque).

Moment of inertia (Wk^2) is often referred to as the flywheel effect and is the product of the weight of the object times its radius of gyration squared. The units are in lb-ft². A centrifugal pump starting with its discharge valve closed typically is a load that can be brought to speed quickly. But large fans are usually high-inertia items whose starting time may be around 20 sec. Cold air imposes a heavier load

than does warm air. Thus, fans that are part of HVAC systems are particularly the type of loads that must be cycled with care.

Besides heating due to the length of time it takes to reach operating speed, motors must be able to absorb and dissipate the energy required to rotate the load. Motors are energy-conversion devices. They consume electrical energy; heat is the result. The energy to be absorbed during an acceleration from rest to full speed with zero load torque is equal to the kinetic energy of the rotating mass at full speed and is independent of starting time. So there are many factors that enter into the determination of permissible starts per hour.

NEMA calculation

Fig. 1 shows how Table 2-2 in NEMA's MG-10 is arranged. Depend-

far exceed them), the value B must be divided by the actual value of Wk^2 for the load of the motor being analyzed. The *lesser* of the two numbers—either the A value or the calculated result of B/Wk^2—represents the maximum number of times per hour that the motor can be started.

Having found the permitted number of starts does not mean that the motor can be started that many times in rapid succession. There must be a *minimum* waiting time between each start to allow the motor to cool sufficiently before the next start. This minimum rest time, in seconds, is given by the value in the appropriate C column.

As an example of using the table, assume that a Design B motor is rated at 1 hp, 1800 rpm, and is directly connected to a pump having a Wk^2 of 0.24 lb-ft². From the table, A = 30 and B = 5.8; B/Wk^2 = 5.8/0.24 = 24.17.

HP	2-pole			4-pole			6-pole		
	A	B	C	A	B	C	A	B	C
1	15	1.2	75	30	5.8	38	34	15	33
1.5	12.9	1.8	76	25.7	8.6	38	29.1	23	34

FIG. 1. Arrangement of NEMA Standard No. MG-10 Table 2-2, which can be used for calculating the allowable number of starts and minimum time between starts for Design A and Design B motors. Col. A = maximum number of starts permitted; Col. B = Col. A × load moment of inertia (lb-ft²); Col. C = minimum time between starts (sec). See text.

ing upon the number of motor poles (a 3600-rpm 60-Hz motor has 2; 1800 has 4; 1200 has 6), there is given an A, B, and C value for each hp size. The A and B values do not have any relation to whether the motor is NEMA Design A or Design B. The A value indicates the recommended *maximum* number of starts based chiefly on the requirement to minimize the effect of winding stress imposed by repeated starts. Value B is the number of permitted starts (A) multiplied by a value of Wk^2 that industry-wide experience has shown a standard squirrel-cage induction motor of that hp can accelerate without causing injurious temperature rise. Standard conditions of voltage, frequency, temperature rise, etc. were used in deriving these values of Wk^2. Because the loads to which motors are actually applied tend to vary from the recommended Wk^2 limits (fan loads generally

Thus the motor can be started 24 times per hour and must have a minimum of 38 sec (the value shown in column C for the rated hp) between starts.

The information available in the NEMA standard should be used only as a guide. All pertinent factors must go into actual calculations. For example, a belt-driven item imposes a different load on the motor from one with a similar Wk^2 that is directly connected. The ratio of the load and motor rpm must be factored in. Also, ambient temperature significantly affects the number of times a motor can be started without an excessive temperature buildup. The lower the temperature, the more rapidly the heat can be dissipated. Other factors—shock loading, method of deceleration (coasting, mechanical or dynamic braking), reduced-voltage starting—all have different heating effects.

Fire pumps and controllers

Here are the NFPA requirements for these units, some of the NEC rules that apply, and the equipment that is available to meet the needs.

SAFETY is increased and fire-insurance rates are reduced when a sprinkler system is included in buildings and manufacturing facilities. Fire pumps, either electric-motor-driven or engine-driven, are integral parts of the fire-protection system. They are used where the water requirements, which can be in the thousands of gallons per minute range at high pressures, cannot be met by public supplies or gravity tanks. It is important to understand what type of equipment and starting and control devices are suitable for this service.

National Fire Protection Association (NFPA) Standard NFPA-20 is the key document for the selection and installation of fire pumps and their controllers. Unlike most other equipment, fire-pump controllers are required by the standard to be purchased under a unit contract along with the pump and driving means. This establishes single-source responsibility for the satisfactory operation of the *entire* fire-pump system. Other NFPA rules define the requirements for fire-pump motors, controllers, installation, and testing and reference NEC sections that detail protection, feeder sizing, etc.

FIRE PUMPS AND FIRE-PUMP CONTROLLERS are specialized to their tasks. Both the driver and controller must conform to the requirements of NFPA-20, *Centrifugal Fire Pumps*. The jockey pump controller on the left of the fire-pump controller is used for equipment that is used to maintain a specific pressure within fire lines. Jockey pumps make up for leaks in the lines. Their controllers must also conform to NFPA-20.

Motors

Squirrel-cage and wound-rotor AC motors or DC motors are permitted by NFPA-20 to drive fire pumps. Normally it is the squirrel-cage induction motor that will be quoted when bids are solicited for motor-driven fire pumps. If there are specific requirements for other types of motors, these must be spelled out in the equipment specifications.

The NFPA standard places responsibility for providing a properly sized motor for the driven equipment with the pump manufacturer. The motor must have a full-load amp (FLA) rating that will not be exceeded under any conditions of pump load at the rated voltage and frequency of the motor. However, this does not mean that the electrical specifier has no input. If the motor has a service factor stamped on its nameplate, the FLA rating is increased to that permitted by the service factor. Thus the manufacturer can utilize this service factor in meeting the requirement. If a service factor is required by the purchaser for future use, this fact must be made known to the vendor.

Other items that must be included in the specs are: the required motor enclosure (open, dripproof, TEFC, explosionproof, etc.); the altitude at which the equipment is to be installed if over 3300 ft; and the existence of unusual moisture or abrasive-dust conditions.

Motor nameplate markings, bearing types, and terminal-box design are all detailed for the pump manufacturer in the standard. Instructions for the care and lubrication of motor bearings are required to be supplied with each motor.

When electric motors are used in a high-rise building whose height is above the pumping capacity of the fire department, the NFPA requires that a reliable emergency source of power be provided such as an engine-driven generator, or that standby engine-driven fire pumps be included in the fire-protection system. The emergency generator must be sized adequately to allow starting and running of the fire pumps as well as any other essential services connected to the generator.

Electric drive controllers

All controllers must be specifically listed for use as part of a fire-protection system. They include nonautomatic and automatic types used with electric-motor-driven pumps.

Electric drive controllers are available for across-the-line and reduced-voltage service. Primary-resistor, primary-reactor and wye-delta reduced-

FIRE-PROTECTION SYSTEMS can be very extensive and sophisticated. Here research-assistant Wendy Hutchinson programs the computer-operated system installed at the Battery Energy Storage Test facility located in Hillsborough Township, NJ. The system will detect the location of a fire and, depending upon its nature, will activate the water deluge, sprinkler, Halon gas, or dry chemical systems.

voltage types can be specified. In addition, there are listed starters for jockey pumps and for limited-service applications. Jockey pumps are used as part of a fire-protection system to compensate for minor leaks. NFPA-20 prohibits fire pumps from being used to maintain pressure. Limited-service pumps are used to provide some additional fire protection where none is mandated by building codes. They are limited to applications where the fire-pump motor is 30 hp or less and only as permitted by the authority having jurisdiction.

Controllers must have a short-cir-

cuit-current withstand rating at least equal to the available short-circuit current for the circuit in which it is to be used. A table is included in NFPA-20 for determining the short-circuit current at the controller for some limited cases. Where the table does not apply, the value must be determined by a short-circuit study. The standard prohibits the purchase of centrifugal fire pumps until all conditions under which they are to be installed and used are determined. This can be interpreted to include the data necessary to make a short-circuit evaluation.

Details of the requirements of the isolating switch, circuit breaker, motor starter, mandatory alarms, drawings, markings, etc. are all included in NFPA-20. The entire unit must be assembled, wired and tested by the manufacturer. Once delivered, it becomes the responsibility of the pump manufacturer to properly integrate the various components into a package. Also, the pump manufacturer or a representative is responsible for having the controller's service personnel on the site for adjustments, repair during the warranty period, and for testing after installation.

Those designing the fire-protection system and specifying the controller have a significant role to play in the selection process. If it is known that the fire pump is located in an unattended area, remote alarms and signals *must* be provided at a point of constant attendance. A source of separate reliable voltage (125 V or less) must energize alarms indicating loss of line power and fire-pump motor running conditions. This need must be included in the specs because the controller manufacturer must provide the proper contacts for actuating the alarms. To avoid possible misunderstanding with the *authority having jurisdiction* who approves the installation when completed, controller manufacturers can provide alarm panels that have been approved for installation in remote locations.

The choice between automatic and nonautomatic control of the fire-pump motors depends upon the protection system design. However, the automatic controller—which is normally started by a pressure switch sensing low line pressure—must also be manually operable. The controller manufacturer will provide the needed equipment to satisfy the standard's requirements when either of the methods is selected. How-

ever, if some other automatic means is included in the fire-protection system—such as the operation of a deluge valve—this fact must be mentioned in the spec so that the controller vendor can accept this remote start signal. The signal must be in the form of a normally closed contact on the fire-protection system.

When more than one motor-driven fire pump is included in the system, the pumps must be started in sequence with intervals of 5-10 sec between them. The manufacturer must be aware of this so that the proper timing relays can be included in the controller.

Manual or automatic shutdown of the pumps can be specified. When only one pump is used and it is the sole supply of the system, *manual* shutdown after automatic start is mandatory. When automatic shutdown after automatic start is specified, timers must be included in the controller to allow the pumps to run for a set period of time after the automatic start conditions have corrected themselves.

Diesel engine drives and controllers

Engine drive selection for these types of fire pumps requires little electrical input. All engines listed for fire-pump service must have the starting and monitoring devices required by NFPA-20, with all components wired to the terminals in an engine-mounted junction box. The terminal-block numbering must match those of the engine's controller. However, selection is required of the two batteries that are mandated for an electric-start engine. The choice is between lead-acid or nickel-cadmium batteries and whether the voltage is to be 6 or 12 V.

Specifications for engine-driven fire-pump controllers should include a list of options that are desired beyond the standard items required by NFPA-20. The most critical item is whether the battery charger should be part of the controller. Because of specific monitoring and sequencing requirements mandated by the standard, integration into the controller is preferable but is generally offered only as an option. There are specific requirements for certain pilot lights and alarms that must be supplied with the equipment, but units that are to be mounted in unattended locations must have remote signals and alarms at a point of constant attendance. Listed panels that meet the

AUTOMATIC TRANSFER SWITCHES must be a part of a controller that is used in an installation that requires more than one source of power. The alternate sources can be either a utility line, an in-plant generating system, or an emergency generator. When the fire pump is motor-driven, an alternate diesel-engine-driven fire pump can be installed as a backup source of water.

needs of engine-driven fire-pump installations are available. The controller manufacturer must be aware of remote indication or operation needs. In addition, the requirement for timers needed for sequence starting of protection systems that include more than one engine-driven pump must be listed in the specs.

A weekly program timer must be part of an approved controller. Its purpose is to start and exercise the engine automatically at preset times. An often-asked-for option is a pressure recorder mounted in the controller. This assures that a record of engine startup is available. Many other such options are usually offered by controller vendors for convenience of operation, maintenance, or to meet requirements set by the authority having jurisdiction for approving a fire-protection system.

Some design and installation considerations

Fire pumps are exempted from several NEC rules that apply to electrical installations. For example: they may

be connected to the *supply side* of the service disconnect of the facility (Sec. 230-82, Ex. 5). Alternatively, Sec. 230-2, Ex. 1 permits fire pumps to be fed from a service separate from that of the building. In this case, a permanent plaque must be mounted at each service-entrance location that identifies the other location and the extent of the electrical system it feeds. When the fire-pump controller acts as the service-entrance equipment, it must be listed for that purpose.

Reliability of the service to the fire pump is very important. The utility line or an onsite generation facility must be designed and arranged so that there is little or no chance of interruption of the service. If this is not possible, the fire pump must be fed from both sources or from either of them along with an emergency generator. The two sources should be run by separate routes or in such a way that dual failure at one time is only a remote possibility. Underground routing is particularly recommended. An *automatic* transfer switch located in the fire-pump room must be included in the design. NFPA-20 requirements for the design and application of the switch must be followed.

The feeders to the fire pump must not be routed through other buildings. The "outside the building" requirement can be satisfied by enclosing the circuit conductors in 2 in. of concrete or an equivalent 1-hr fire-resistance-rated protective arrangement.

NFPA-20 does not recommend that voltages above 600 V be used for fire pumps. Where this limitation is impractical, the use of higher voltage may be acceptable (to the authority having jurisdiction). Receiving this waiver may be difficult because HV controllers for fire-pump service are not listed. This means that in many cases the distribution voltages within a facility must be stepped down. This makes the NFPA rule *against* having a disconnecting means as part the fire feeder circuit difficult to apply rigidly. Sizing of feeders must be per the applicable NEC Article 430 rules.

Fire-pump motors and their controllers must be installed so that the current-carrying parts are a minimum of 12 in. above the floor. Batteries required to start diesel-driven fire pumps must also be mounted on racks at an elevation that will prevent them from being flooded by water.

The controller must be located as

SENSORS AND EXTERNAL WIRING of fire-pump controllers is required by NFPA-20 to be protected from damage. Wiring must be run in rigid metal or liquidtight flexible metal conduit. Shorted, cut or damaged external control components must not prevent the fire pumps from starting.

close as possible to and within sight of the motor or engine. NEC Article 100 defines "within sight of" as being visible and not more than 50 ft distant.

Suitable means must be provided to maintain the temperature of a fire-pump house or room above 40°. This minimum requirement may have to be increased if the driver is a diesel engine. The manufacturer's recommendations on room temperature and the need for water and oil temperatures must be followed. Ventilation, lighting, and fixed or portable battery-operated emergency lights must also be provided.

Fire-pump control circuits are critical items. NEC Article 430, Part F, thus exempts them from the requirements for overcurrent protection of the wiring and control transformers. Circuits that extend beyond the controller must be designed so that breakage or shorting of wire and equipment or loss of power may start the pump, but in no case must it prevent the units from starting. The wiring must also be protected against mechanical injury by being installed within rigid or liquidtight flexible conduit.

An additional item should be noted. The 1983 version of NFPA-20 has dropped a provision that previously had permitted the continued operation of existing fire pumps that complied with the NFPA standard applicable at the time they were built.

PART III

TROUBLESHOOTING and MAINTENANCE

Why motors fail

By RICHARD SCHEINERT

WHEN A MOTOR comes into our plant for repair, we inspect it thoroughly, searching carefully for the possible cause of failure. Pinning down the exact cause is not easy; often the source of trouble has been obscured by burned windings or other misleading "faults." For example, the motor windings may be badly burned; but closer examination may reveal a damaged ball-bearing that could have caused the rotor to rub on the stator windings. Taking this analytical thinking process a little farther—why did the bearing fail? Was it misalignment, excessive loading, or simply a lack of proper lubrication?

Many motor failures can be averted, or at least the useful life of a motor can be extended, if proper preventive measures are taken. And an important part of this process is knowing why motors fail.

We have found that sources of motor troubles generally fall into one of the following categories:

(1) Harsh environment

(2) Improper motor selection/application

(3) Inadequate installation

(4) Mechanical failure

(5) Electrical problems

(6) Voltage unbalance

(7) Inadequate maintenance

(8) Combination of one or more of the above.

Harsh environment—Frequently, excessive temperature (whether it is caused by the environment or by a problem within the motor) is a cause of motor failure. Motors must operate within their nameplate temperature-rise rating to assure long motor life. For every 10 degrees a motor is operated above its rating, the insulaton life

MOTORS BEING REBUILT in Scheinert plant are first carefully inspected and tested; then any pertinent information is analyzed to find cause of failure.

RICHARD SCHEINERT is president of R. Scheinert & Sons, Inc. Electro-Mechanical Service, Inc., Philadelphia, PA.

of the motor is halved.

In addition to maintaining proper ambient temperature, other sources of elevated temperature must be found and eliminated. These sources could include misalignment, overload, incorrect voltage, and many others. Other harmful environments include those containing corrosive fumes or vapors, salt-laden air, and excessive dirt, dust, or other contamination. In these locations, motor enclosures designed for the surrounding atmosphere or conditions are essential.

Moisture is another common source of motor failure. If moisture forms on the surface of the insulation because of high humidity, temperature change, or direct exposure to water, the insulation surface may become highly conductive and result in an insulation failure and immediate motor failure. Also, it is possible for the insulation to absorb moisture over a period of time until the insulation resistance becomes so low that the fault occurs.

Improper motor selection/application—The variations and degree of incorrect selection and application vary greatly. Sometimes the misapplication is so minor that the motor survives for a significant time. It is essential that the proper size and type of motor be selected for the load. The motor manufacturer, your local service firm, and motor standards can provide guidance. There are numerous factors to consider. As an example, a severe duty cycle could cause premature motor failure. Jogging, plugging, and long acceleration cause motors to run at lower-than-normal speeds. Because they draw high currents when starting, motors subject to these duty cycles will often experience excessive heating because of the higher currents. Also, normal cooling derived from the revolving rotor will be greatly diminished, adding to the overheating problem.

Consideration of altitude is another important application factor often overlooked. At high altitudes, the lower air density reduces the effectiveness of cooling air. This reduced cooling permits the running temperature of most motors to increase approximately 5% for each 1000 ft of elevation.

Selection of the proper motor enclosure is important. Standard motor enclosures are available for practically any situation.

Inadequate installation—Faults in motor mounting can cause motors to fail. If mounting bolts are not tightened or not sized properly, misalignment or vibrations occur, resulting in bearing and shaft failures and eventual winding burnout. Mounting steel, foundations and grouting must be strong enough to withstand starts and stops.

Couplings, belts, sheaves and any other connections between the motor and driven load must be aligned correctly to avoid excessive vibration that is so damaging to motors.

Also, inadequate installation occurs when the proper attention is not given to the NEC and other standard installation practices. Article 430 of the NEC and NEMA standards provide installation guidance.

Mechanical failure—An excessive load can quickly result in motor failure. The motor may have been properly sized for its load initially; however, a change in the load, or in the connecting drive, could result in overloading of the motor. Bearings may begin to fail, gears may begin to bind, or any source of additional friction or load can occur. When this happens, the motor will draw additional current, resulting in increased motor temperature. If the motor current exceeds rated full-load current for any length of time, the overheating will shorten motor life. Very high overcurrents should trip the overload relays if they have been properly sized.

Bearing failures are one of the most common causes of motor failures. It has been estimated that up to one-half of all motor burnouts have been caused by a bad bearing. The many causes of bearing failures and techniques to properly maintain bearings should be thoroughly understood.

Misalignment of motor and driven load, couplings, gears, pulleys, and belts are all mechanical causes of motor failure. All components should be dynamically balanced to assure best possible conditions for long motor life. This will, in addition, minimize vibration and associated problems.

Electrical problems—Incorrect supply voltage to a motor will shorten life and could cause rapid failure if voltage deviation is excessive. Low voltage will cause higher-than-normal current. If the voltage decrease is severe enough, the excessive current will cause the motor to overheat.

A high line voltage to the motor reduces copper losses; but the increased magnetic flux results in higher iron losses. A small increase in supply voltage could reduce the current drawn by the motor; however, higher increases—on the order of 10% or more over nameplate voltage—will cause saturation of the iron and a significant increase in current with the accompanying harmful overheating of the motor.

Unbalanced voltages—Unbalanced 3-phase voltages to a motor can result in a very high current unbalance. Very high motor currents can occur, resulting in fast overheating. Protection must be provided against this kind of problem, and overload relays usually can do the job. Recently, new types of overload relays have been successfully applied to protect the motor not only from unbalanced voltages but also from single phasing—which in actuality is an extreme form of voltage unbalance.

Inadequate maintenance—Frequently, basic preventive maintenance could prevent or at least delay the failure of a motor. When our troubleshooters have visited certain locations, they report conditions such as dust and dirt on motors or clogged ventilation passages, hot-running motors, incorrect motor currents, noisy bearings, moisture in the area of the motor—all of which indicate a lack of planned regular maintenance.

Certainly not all motors need or deserve preventive maintenance, particularly where the maintenance costs would exceed the cost of that motor failure. On the other hand, where the motor is applied in a critical application or is large, expensive or hard to replace, then a preventive maintenance program is well justified. We have made such surveys for industrial plants and, as a result, their production is more reliable, motors last longer, and total operating costs are lower.

VOLTAGE AND FREQUENCY VARIATIONS

Induction motors will operate successfully under running conditions at rated load with a voltage variation of ±10% at the motor terminals. A frequency variation of ±5% is permissible. A combined variation of voltage and frequency of ±10% is acceptable, provided the frequency variation does not exceed ±5% of rated frequency. For example, the voltage could vary by a maximum of ±7% and the frequency by a maximum of ±3%. The combined variation does not exceed ±10%. However, some loss in performance occurs when a motor is operated at other than rated values.

A 5% increase in frequency will increase the speed of the motor about

Characteristic	Undervoltage	Overvoltage
Starting amps	Increase	Decrease
Full-load amps	Increase	Increase
Efficiency	Decrease	Decrease
Power factor	Increase	Decrease
Torque	Increase	Decrease

5% and will increase the efficiency, power factor, and full-load current slightly. Torque will drop about 10%. A 5% decrease will increase the torque about 11%, decrease the speed about

5%, and slightly decrease the efficiency, PF and current.

The accompanying chart shows how motor characteristics are generally affected by voltage deviations.

Anatomy of an all-too-common disaster:

Motor burnout on single-phasing

Loss of three 100-hp motors in an industrial plant emphasizes that properly set conventional running-overload protection in controllers for 3-phase motors is no assurance of protection against costly motor burnouts on "single-phasing" of the supply circuit.

IN A MODERN, well-designed and effectively maintained electrical system in an industrial plant, a recent case of costly motor damage due to "single-phasing" of the supply system underscores a painful reality of the electrical industry—namely, that constantly growing motor applications in all kinds of electrical systems all over the nation is being paced by a constantly growing experience with motor burnouts due to "single-phasing."

Although the National Electrical Code requires protection for motors against running overloads, there is no mention at all in the NEC of any need for protection against single-phasing damage—the damage that can be produced in 3-phase motors when one of the three phase legs of the motor circuit loses voltage due to blowing of one of the three fuses protecting either the branch circuit to the motor or the feeder or service supplying the branch circuit. Three running-overload relays in the motor starter can protect against motor damage by opening the starter under certain conditions of single-phasing, but the hard truth is that three overload relays constitute only partial protection under very specific conditions of motor loading and application.

Complete and effective protection against the threat of single-phasing damage is an extremely important task for the motor-circuit designer. Some motor controller manufacturers incorporate anti-single-phasing in their equipment (in particular, the European designs), but the majority of 3-phase motor starters do not include such pro-

tection. As described in the following case study of single-phasing burnout, existing starters can be retrofitted with the necessary electrical components to provide such protection—although such an addition, as in the case here, may be a case of locking the barn door after the horses have been stolen.

As shown in Fig. 1, the single-phasing damage in the industrial plant involved three of four 100-hp motors driving ammonia compressors. One 100-hp motor and three other smaller compressor motors were subjected to the same single-phasing as the burned-out motors, but the running overload relays in their starters opened the

starters and saved the motors from damage. The single-phasing occurred as a result of blowing of one of the three 3000-A silver-sand current-limiting fuses in the main service disconnect of the plant's main switchboard. A very important factor in this incident is that the single-phasing occurred during a utility brownout (reduced voltage) late in the afternoon of a hot summer day.

The following are excerpts from the engineering department's report on the incident:

At 5:02 p.m. one of the main fuses in our main service disconnect blew, causing single-phasing of the system, which resulted in the ultimate loss of three

Fig. 1. Three of the four 100-hp motors burned out due to single-phasing

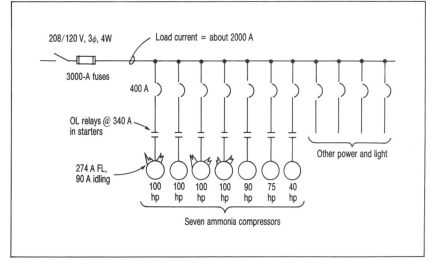

208/120 V, 3φ, 4W

Load current = about 2000 A

3000-A fuses

400 A

OL relays @ 340 A in starters

Other power and light

274 A FL, 90 A idling

100 hp 100 hp 100 hp 100 hp 90 hp 75 hp 40 hp

Seven ammonia compressors

FIG. 2. Main switchboard contains the 3000-A fused main service disconnect switch (left arrow) and the 400-A CBs for the 100-hp compressor motors as well as the branch-circuit CBs (right arrows) for other equipment, including the smaller compressors at left (90, 75 and 40 hp).

100-hp motors. Two were total burnouts requiring replacement, and one was severely damaged but capable of being rewound. When the single-phasing occurred, plant personnel were immediately alerted to a problem because of immediate loss of lighting fed by the opened phase. Within 30 minutes, smoke was noticed at the motors, which continued to "cook" until they burned their supply open. Neither the motor starters nor the branch circuit breakers operated to clear the faults.

Two of the motors (one burned-out, one damaged) were driving ammonia compressors for process line No. 1; the other burned-out motor was driving an ammonia compressor for process line No. 2. At that time, line No. 2 was already operating on its second shift, and line No. 1 was preparing to commence running. We believe all compres-

FIG. 3. This is the main 3000-A current-limiting (silver-sand) fuse that blew and caused single-phasing of the whole system.

sors for line No. 1 were turned on but operated unloaded. The same was true of the damaged motor on line No. 2.

When the single-phasing occurred by blowing of one of the main 3000-A fuses, the maximum service current draw was probably not over 2000 A—in service fuses that are rated at 3000 A. The power company's pole-line protection did not trip, and there is no other indication of

any other overload that would cause the one fuse to go, although there was a brief high-voltage transient (microseconds in duration) passing through our switchboard from the utility lines. The main fuses in the switchboard main were originals dating back to 1967, and the blown fuse may have metal-fatigued and crystallized due to high-heat cycling over the many years of operation.

Because compressor motors, in particular, run in both full-load and unloaded conditions, they are extremely vulnerable to burn-out if they single-phase while in an unloaded condition. Anti-single-phasing protection is being added to all of the compressor starters. This is a protector that will drop a motor out if one of its phases loses current flow. There is nothing that could have been done to protect against this failure other than to have put these phase protectors in earlier.

Fig. 2 shows the main service switchboard that contained the 3000-A main fuse that blew (in main disconnect section at left arrow). The 400-A CBs providing motor branch-circuit protection for the 100-hp motors are shown (right arrows) in the board and were not actuated by the single-phase fault condition. Fig. 3 shows the main 300-A fuse that blew and caused the single-phasing. It was hypothesized that one or more of its internal links has opened over the years, derating the fuse continuous-current value to something less than 3000 A. In such a condition, the fuse would be susceptible to blow on an inrush current that it was exposed to while preheated to a high temperature by the fairly constant demand load current of around 2000 A. Discoloration of the fuse blades indicated poor contact between the bolted blades and the fuseholders in the main switch, which could have made a significant contribution to fuse overheating and utlimate blowing.

Fig. 4 shows one of the replaced 100-hp motors, which is controlled by the part-winding starter shown in Fig. 5. After the single-phasing damage to the motors, the starter for each compressor motor was equipped with anti-single-phasing protection, as shown in Fig. 5 and the close-up of Fig 6. The current transformers and the relay shown at left provide for opening the starter on failure (opening) of any one of the three phases. Such a relay hook-up can also protect against unbalance in phase currents, such as might occur due to a partial winding short or a ground fault in a motor winding. Sensing such fault early and opening the starter can reduce the extent of damage to a motor.

Single-phasing action

Fig. 7 shows the typical characteristics of a squirrel-cage induction motor as related to a 100-hp unit. At no-load (or light-load), both efficiency and

FIG. 4 This is the replacement motor for one of the three 100-hp motors that were completely destroyed. A belt drive from the motor shaft (rear) drives the compressor clutch.

FIG. 5 Part-winding starter for each dual-voltage 100-hp motor was retrofitted with phase-failure protection (arrow).

FIG. 6 Solid-state phase-loss/phase-reversal relay (arrow) is connected to external current-transformers on the three conductors of the motor circuit.

power factor are drastically reduced from their full-load values, and that can account for very high no-load motor currents during single-phase operation. The characteristics of the graph are applied to analysis of the basic concepts behind single-phasing of a 3-phase motor.

Analyzed by symmetrical components, a single-phasing condition involves two balanced polyphase systems of opposite phase sequence—producing a positive and negative torque with the resultant effective torque being the difference between the two. Depending upon the type of motor winding (star- or delta-connected), some amount of harmonic current is generated in the windings and contributes to exagger-

Fig. 7. Typical variation of electrical characteristics of a 3-phase, squirrel-cage induction motor with variation in loading

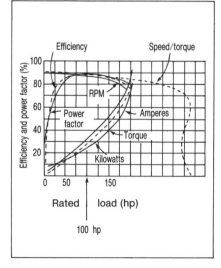

ated heating effects within the windings under single-phasing conditions.

Fig. 8 presents a simplified comparison between normal 3-phase operation of a fully loaded 100-hp motor and single-phase fault operation of the same motor fully loaded. The analysis is intended to illustrate the difference in the general nature of each type of operation. The numerical values are selected as typical and representative of the changing actions from normal to single-phase operation. Change in power factor and efficiency accompany the change in operating conditions as a result of shifts in the vector relationships between currents and voltages. For practical purposes, it can generally be assumed that the single-phase current under full-load conditions is about equal to 1.73 (the square-root-of-3 factor from 3-phase calculations) times

Fig. 8. Conventional OL relays can protect against single-phasing damage to fully loaded motor.

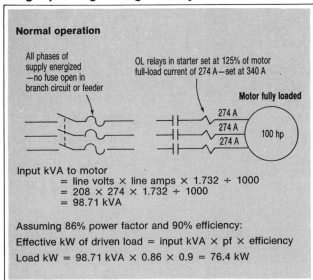

Normal operation

All phases of supply energized —no fuse open in branch circuit or feeder

OL relays in starter set at 125% of motor full-load current of 274 A—set at 340 A

Motor fully loaded

274 A
274 A
274 A

100 hp

Input kVA to motor
= line volts × line amps × 1.732 ÷ 1000
= 208 × 274 × 1.732 ÷ 1000
= 98.71 kVA

Assuming 86% power factor and 90% efficiency:

Effective kW of driven load = input kVA × pf × efficiency

Load kW = 98.71 kVA × 0.86 × 0.9 = 76.4 kW

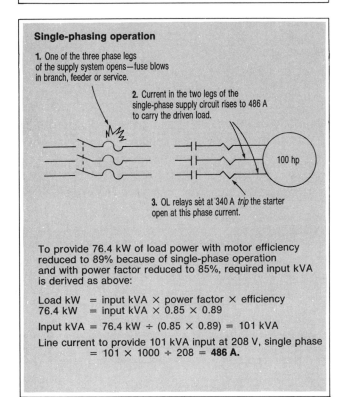

Single-phasing operation

1. One of the three phase legs of the supply system opens—fuse blows in branch, feeder or service.

2. Current in the two legs of the single-phase supply circuit rises to 486 A to carry the driven load.

100 hp

3. OL relays set at 340 A *trip* the starter open at this phase current.

To provide 76.4 kW of load power with motor efficiency reduced to 89% because of single-phase operation and with power factor reduced to 85%, required input kVA is derived as above:

Load kW = input kVA × power factor × efficiency
76.4 kW = input kVA × 0.85 × 0.89

Input kVA = 76.4 kW ÷ (0.85 × 0.89) = 101 kVA

Line current to provide 101 kVA input at 208 V, single phase
= 101 × 1000 ÷ 208 = **486 A.**

Fig. 9. OL relays may not protect unloaded or lightly loaded motor against single-phasing

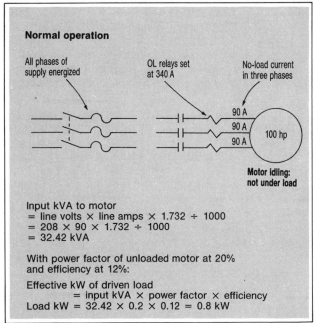

Normal operation

All phases of supply energized

OL relays set at 340 A

No-load current in three phases

90 A
90 A
90 A

100 hp

Motor idling: not under load

Input kVA to motor
= line volts × line amps × 1.732 ÷ 1000
= 208 × 90 × 1.732 ÷ 1000
= 32.42 kVA

With power factor of unloaded motor at 20% and efficiency at 12%:

Effective kW of driven load
        = input kVA × power factor × efficiency
Load kW = 32.42 × 0.2 × 0.12 = 0.8 kW

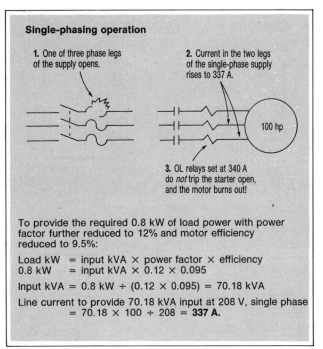

Single-phasing operation

1. One of three phase legs of the supply opens.

2. Current in the two legs of the single-phase supply rises to 337 A.

100 hp

3. OL relays set at 340 A do *not* trip the starter open, and the motor burns out!

To provide the required 0.8 kW of load power with power factor further reduced to 12% and motor efficiency reduced to 9.5%:

Load kW = input kVA × power factor × efficiency
0.8 kW   = input kVA × 0.12 × 0.095

Input kVA = 0.8 kW ÷ (0.12 × 0.095) = 70.18 kVA

Line current to provide 70.18 kVA input at 208 V, single phase
= 70.18 × 100 ÷ 208 = **337 A.**

the normal 3-phase full-load current.

When the same 100-hp motor is operating unloaded, the line current is only 90 A, as shown for normal operation in Fig. 9—with the power factor and efficiency at very low value (as indicted in Fig. 7 for no-load condition). The effective kW rating of the load represented by the motor and clutch—the basic no-load condition—is only 0.8 kW. As shown in the bottom of Fig. 9, power factor and efficiency

change substantially under single-phasing conditions—where shifts in PF and efficiency occur along the almost vertical curves for power factor and efficiency at extremely low loading (left bottom of Fig. 7). Under such conditions, it is readily possible to produce motor currents in excess of the full-load value of 274 A but below the actual trip setting of the OL relays. As shown, a current of 337 A could flow continuously.

The value of 337 A is meant only to be indicative of the elevated current that might exist. In fact, if particular OL heaters happen to be actually rated at the high side of the manufacturing tolerance band for the trip current, even higher continuous current would be possible. In addition, there is always the possibility that the time-current characteristics of the heaters would permit excessive currents (above the heater pickup value) for a long enough time to damage the motor before the relay opens the starter. Fig. 10 shows the minimum-maximum operating band for one manufacturer's overload relays, with the maximum trip curve allowing 150% of motor full-load current for up to 10,000 seconds, which is over two hours and forty-five minutes—enough time to do serious damage.

Protection against damage due to single-phasing can be facilitated by careful selection of the type and rating of running overload devices in motor starters. Whether an overload relay is a bimetallic or solder-pot thermal device or is a magnetic type of device, the shape of the inverse-time-current characteristic should conform as closely as possible to the motor's heat damage curve (the time-current curve defining the limits of motor heating

damage). Selection of overload relays is a more critical task than it used to be because of the wide variety of types and characteristics of both electromechanical and solid-state relays available today from the broad range of manufacturers.

Several manufacturers incorporate phase unbalance and phase failure (anti-single-phasing) protection in their line of motor starters. These include overload relays that trip faster and at a lower current when single-phasing occurs. But anti-single-phasing protection is more important for motors rated above 20 hp, in which sizes the differences between no-load current and full-load current are much greater than in smaller motors, thereby taxing the ability of conventional OL relays to protect against overcurrnt at any conditon of loading. For larger motor applications, the starter can be equipped with a 3-phase power monitor of the type shown in Fig. 11. In that hookup, the relay connected into the control and power circuits will drop the starter out on loss of any phase, on low voltage on any or all phases, or even phase reversal. The unit consists of a solid-state circuit hookup that senses voltage and phase angle and controls a double-pole, double-throw relay as shown.

Fig. 11. Starter can be equipped with a sensing and relay assembly that trips out starter and provides alarm on single-phasing of the motor supply circuit.

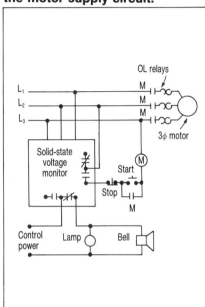

The bottom line on all of this is that costly motor damage due to single-phasing is today, more than ever before, a serious threat that must be anticipated and eliminated.

Fig. 10. OL protection may not act fast enough on low-end overcurrent

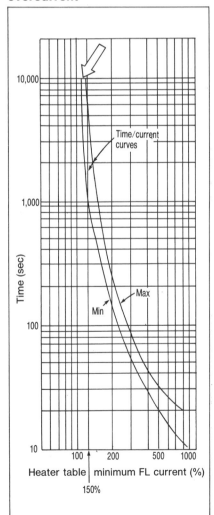

COSTLY MOTOR REPAIRS, rewinding, or replacement can be avoided by taking adequate precautions against damage caused by "single-phasing" of the supply circuit.

Motor thermal protection minimizes downtime

Sensors embedded in motor windings and placed in contact with bearing races reduce motor burnouts, eliminate nuisance tripping, extend motor life, and minimize equipment shutdowns. *

THERMAL SENSORS (foreground) are used in conjunction with control module shown to provide accurate overtemperature protection of standard 3-phase motors.

WINDING burnouts, nuisance shutdowns, and bearing failures are among the major motor problems which plague industry with frequent reoccurrence. In most instances, these can be traced to conditions resulting from improper motor control and operation. This need not occur; inherent thermal protection systems are available which can safely and accurately limit motor temperatures to acceptable levels even under abnormal operating conditions.

Motor failures cause costly, unscheduled equipment downtime. With material and labor costs on the rise, it becomes mandatory that all available means be used to keep equipment operating at maximum efficiency. Downtime production losses on a typical automatic transfer line have been known to cost $1500 per minute, excluding normal maintenance costs for repair of the fault. Even in less critical operations, the cost of equipment downtime becomes prohibitive when extended beyond an hour or two.

Downtime periods are of two general types: those that occur due to motor failure such as bearing seizure or motor winding burnout, and those resulting from temporary shutdown

*Based on a paper by Robert E. Obenhaus, Engineering Supervisor, Motor Controls Dept., Texas Instruments Inc., presented at the 25th National Plant Engineering and Maintenance Conference in March, 1974 in Cleveland, Ohio.

due to operation of a protection system. Although both occur with regularity in many industries, they are not always systematically recorded to provide meaningful comparisons. For example, motor winding burnout usually is more noteworthy, since a machine operation is stopped and several people become involved in restoring the equipment to proper running order. In these instances, documentation is usually more complete regarding the nature of the fault, its correction, and subsequent production loss.

This is not necessarily true where equipment is shut down due to operation of a protective device. Even though equipment damage generally is avoided or postponed, repeated downtime over a period of months can actually amount to a greater total than for motor burnout conditions. Because these shorter downtime periods are not normally recorded, their cumulative total is not always correctly assessed by management. Furthermore, much of this downtime is of the nuisance variety and could be avoided with proper protection systems. It follows that more effective motor protection could provide a solution to both classes of equipment downtime.

Why motors overheat

Common causes of motor overheating include excessive overloading, bearing seizure and misa-

lignment. However, there are several less dramatic effects which contribute heavily toward motor failure. These include:

1. Slight but sustained running overloads
2. Rapid load fluctuations (jogging and plugging)
3. Restricted ventilation
4. Locked rotor
5. Single phasing on three-phase systems
6. High ambient temperature
7. Excessive duty cycles
8. Excessive load inertia
9. Improper motor-load match
10. High altitude
11. Power supply variations (high or low voltage, unbalanced voltage)

In all of these instances, the end result is higher-than-normal winding temperatures which, if continued, will result in reduced motor life. A further complication is the trend toward reduction in size and lowering of cost in motor design. Because motor manufacturers are utilizing superior insulating materials, and because motors themselves are being allowed to operate at higher temperature limits, the margin of safety is being narrowed. Because the heart of any motor is its winding, excessive temperatures in this region can promote rapid and irreversible damage. As a rule of thumb, it has been determined that for approximately every 10-deg C increase in winding temperature, the useful life is halved. Thus, it becomes essential to limit the winding temperature to safe levels during abnormal operating conditions to assure full motor life.

Motor protection criteria

Section 430 of the NE Code provides for motor and branch-circuit protection. Article 430-32(a)(1) describes a system using motor current as a means of providing an analog of the motor's temperature. Fig. 1 illustrates this system in pictorial fashion.

Although overcurrent systems of various types have been used for many years, their analog nature is not capable of accurately reproducing the temperature conditions of today's modern motors. In most instances an adverse change in a motor's temperature condition is caused by increased current. If an overcurrent device is used, this change is interpreted as an increase in temperature, and the motor is shut down at some point in time.

However, not all factors influencing excessive motor temperature are reflected by increased motor current. Notably in this category are restrictions in motor ventilation and increases in altitude. The inability to detect these conditions, especially blocked ventilation, seriously compromises the protection needed to reduce the incidence of motor burnout and equipment downtime.

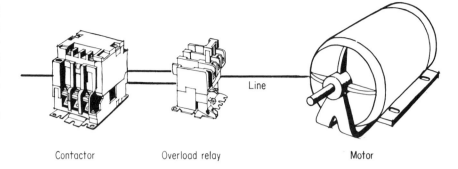

FIG. 1. Conventional system used for overcurrent protection of motors in accordance with Section 430-32(a)(1) of the NE Code.

Heating and cooling rate considerations

A further difficulty encountered with devices which analog motor temperature through current sensing is the matter of different heating and cooling rates between the motor and the protective device. Since a single overload relay with changes in heaters can be used to protect a number of motor sizes, it is practically impossible to achieve a close thermal match on each motor being protected. The motor with the larger thermal mass is prone to heat and cool at rates which are different from the overload relay. Depending upon the nature of the motor's overloaded condition, this characteristic can cause either nuisance tripping or premature motor failure. The former can be caused by slight and sustained running overload resulting in operation of the overcurrent relay before the motor winding reaches abnormal temperatures. The effect of these nuisance trips often promotes arbitrary upward calibration changes in the level of protection to avoid nuisance operation. Premature motor failure can result from repetitive resetting of an over-

load relay under heavy overload or locked-rotor conditions. Because the overload relay's current-sensing mechanism can cool at a faster rate than the motor, it usually can be reset before the motor has had sufficient time to cool.

Effects of ambient temperature

Analog temperature protection by overcurrent means also suffers from differences in ambient temperature between the motor and controller sites. Since thermally activated overload relays are generally affected by their immediate environment, any changes in the ambient temperature due to ventilation or proximity to other controls and heat sources will affect the level of motor protection. Again, depending on the location of the motor and its condition of overload, the protection system can cause either nuisance tripping or premature winding burnout.

An overload relay located in a nonventilated cabinet can cause premature motor shutdown unless suitably compensated. A hotter ambient at the controller site effectively lowers the

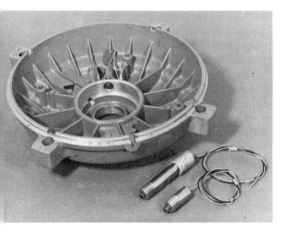

BEARING TEMPERATURE SENSOR is mounted in motor end bell so that spring-loaded tip maintains positive contact with outer bearing race.

FIG. 2. Integral thermal protection system as applied in accordance with Section 430-32(a)(2) of the NE Code.

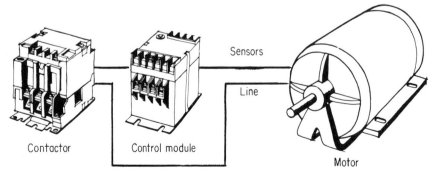

THERMAL SENSOR (arrow) is mechanically tied down and epoxy-applied to provide maximum thermal contact with stator windings. Sensors can be installed easily on either new or rewound motors.

protection point, shutting the motor down prematurely. Conversely, if the motor is located near a source of heat, such as a furnace, the safe temperature could be exceeded before its current reaches the level which activates the overload relay.

Since ambient temperature conditions in control panels vary considerably due to seasonal changes, ventilating effectiveness, and other production related reasons, it is not practical in most cases to compensate these devices every time a change occurs. As a consequence, the degree of overtemperature protection to a motor will be subject to the prevailing ambient conditions.

Integral protection

Section 430-32(a)(2) of the NE Code makes provision for an alternate form of motor and branch-circuit protection—the use of thermal protectors integral with the motor itself. Such protection makes it possible to sense all stator winding overtemperature conditions, including those resulting from blocked ventilation and ambient conditions. One form of this system uses integrally installed thermal sensors in the motor winding with a panel-mounted control module.

In accordance with the provisions of Section 430-32(a)(2), this system, when approved for use with the motor which it protects, also provides protection to the controller and branch-circuit conductors against excessive heating due to motor overloads or failure to start. Fig 2 shows the components of a typical system using thermal sensors both in the motor and in a companion control module. The small size and fast response of these sensors enable them to follow even the higher rates of temperature rise experienced with today's T-frame motors. Bearing-overtemperature sensors are also available and can be used with the same control module.

Of the various types of sensors available, those having abrupt changes in resistance at the temperature trip point offer several advantages. In addition to being available in numerous temperature values, they can be series-connected (up to three) within a motor, requiring just two leads back to the control module. Since the module is not required to make a qualitative judgment on each set of sensors, different temperatures can be used without the need for recalibration. A module with three sensor inputs could, for example, accommodate three motors each protected to different temperature levels. Also, it is possible to intermix series sensors of different temperature set points on a single input channel, since any sensor experiencing a high resistance will trip the control module.

Protection of a motor based on the actual thermal conditions existing within the stator winding is a feature which overcomes many of the problems prevalent with conventional overcurrent systems. Advantages include:

1. Protects against motor winding overtemperature, regardless of the factors responsible for the overtemperature condition.

2. Reacts to motor winding temperature only and is insensitive to temperature conditions at the motor controller.

3. Acts upon total temperature of the winding, factoring in the contribution of ambient to total temperature.

4. Prevents reset of the protective system until there is assurance that motor temperature has dropped to a safe value.

5. Permits remote reset by operating personnel just as soon as safe conditions have been restored at the motor; does not require a qualitative judgment to be made as to whether motor has cooled down sufficiently.

6. Retains an "overload trip" flag for information of troubleshooters when system has been reset by operating personnel.

7. Has built-in, inherent fail-safe protection against short circuit, open circuit and loss of control power.

8. Is tamperproof and cannot have trip point altered by simple adjustment or component replacement at the controller.

9. Utilizes a single, versatile, basic module that can be used to protect most motors of any size in the plant.

10. Can use a single module to handle multiple inputs from several motors if motors are interlocked with each other, lowering "cost per motor" of protection. Each input to the sensing module can be different and matched to the temperature withstand of its specific motor.

11. Can be easily retrofitted to protect most existing motors in a plant.

Most nuisance-type downtime is eliminated with inherent thermal protection systems. Since a motor is taken off the line only when its actual protection level is exceeded, any trip is a "necessary trip" rather than a nuisance one. Because effective locked-rotor protection is achieved by close thermal coupling between the motor winding and the sensing system, the machine operator can be allowed to reset and restart his own equipment safely where faults can be removed easily. The electrician at his option can still investigate the faulted condition— but later, during a noncritical period. Freeing him and other maintenance personnel to work on "real" problems is perhaps one of the greater cost-saving benefits to be realized with inherent thermal protection systems.

Installation

It is relatively easy to add overtemperature protection to most existing industrial motors. It is even easier to install these sensors at the time a motor requires rewinding, before varnish dip-and-bake operations.

The materials and construction of

these sensors make them capable of withstanding the handling and high-temperature bake procedures involved in motor rewinding. Also, the encapsulating materials have been specifically selected to resist deterioration from exposure to common industrial environments, such as heat, varnish, moisture, oil and other contaminants.

Sensors can provide overtemperature protection for both running-overload and locked-rotor conditions. Considering running overload first, the sensor operating temperature should be above the maximum rated motor design temperature and below the maximum allowable winding temperature limit. When sensor temperature is selected close to the maximum normal motor design temperature—taking into account the sensor/system tolerances—motor life is enhanced, and temperatures on locked rotor remain within allowable limits for most induction motors rated 600 volts or less.

Although the sensors can protect for both running-overload and locked-rotor conditions, the existing overcurrent devices should be retained in retrofit applications to ensure compliance with the motor branch-circuit protection requirements of the NEC.

For a typical general-purpose Class B motor, a sensor with a nominal operating temperature of 145C is indicated. This is arrived at by providing for a tolerance of plus or minus 5 deg C for the protection system and a margin of 10 deg C over the maximum design operating temperature of 130C, which is at 1.15 service factor load for this rating. The purpose of this margin is to insure against nuisance tripping of the protection system. At the same time, the maximum sensor (high-side tolerance) limits motor temperature to 150C, which is well below the recognized limit of 165C for running overload. In most retrofit applications, it will be possible to use a single sensor temperature for a given insulation system.

RETROFITTING INHERENT SENSORS*

FORD Motor Co., one of the nation's largest users of integral-horsepower motors, has long felt the need for a motor protection method that would keep overloaded motors on the line as long as possible and which could be reset by the operator without calling an electrician. A pioneer application of inherent temperature sensors was made at Ford's Dearborn plant. It was on a 3-hp, 3-phase, 460-volt motor that had a long history of burnout problems.

Small barium-titanate sensors were attached to the end-turn windings of each of the three phases and connected in series with a relay module. This first unit fully protected the motor under severe locked-rotor conditions amounting to several thousand overload cycles. During the test, the leads were shorted together periodically to evaluate the fail-safe feature of the system under such a condition. In all instances, the motor was shut down when either an overtemperature or a short-circuit condition occurred. Following this, several other troublesome motors were modified to use inherent protection. In all cases, the problem was solved.

This was the start of a new business for apparatus service firms in the Detroit area. Following are observations of two such shops, Jay Electric and A & C Electric, on their experiences with the new form of protection.

Jay Electric

Jay Electric began implanting temperature sensors in motors for Ford Motor Co. plants in the Detroit area about four years ago. At that time it was almost on a developmental basis, and the customer was keeping close tabs on each motor's performance after it went back into service. However, no motor protected by the new system has yet been returned for further repair, and for the past several months virtually every rewind job for

SENSOR TIP (above) is implanted in a crevice in the windings just back from the nose of the coil by Jay Electric personnel. Match-head-size "bead" (right) is added after coils are inserted.

Ford has been accompanied by an order for the sensors.

Al Leach of Jay Electric reports that installation is easy when it's done at the time of rewinding, adding only 15 or 20 minutes to the job. The sensors are positioned in the end turns of one winding from each phase so that they are surrounded by windings.

Jay's practice is to place the sensors around the circumference of the stator windings approximately 120 deg apart. This is more difficult than placing them on adjacent phase coils, because the repairman must count the

coils around the circumference carefully to make sure that one sensor is on each phase. However, Jay's method gives better protection against the effect of a local external heat source which could raise the temperature of one part of a motor more than the remainder. (This could happen, for example, to a motor mounted on a furnace or next to uninsulated hot pipes.)

Jay feels that the ideal place to sense true winding temperature is about halfway between the insulation spaces and halfway between the nose of the bundle and the end of the stator slot. A sensor placed too close to the nose of the coil will sense a lower temperature than the rest of the coil, since it comes into contact with more cooling air. Placed too close to the slot, it may pick up heat from the laminations or housing, which won't necessarily represent winding temperature.

So far, all motors that have been retrofitted at Jay have been the standard form-wound type, but experiments with other types of motors indicate that other types of windings will not create any problems.

A & C Electric

Most of the service work done by A & C Electric Motor Repair Co. is for plants of the big-three ·automobile companies, but they also do work for suppliers to the automotive industry.

A & C began doing implants of sensors for local plants of Ford Motor Co. about two years ago. They found that it takes from 30 to 45 minutes to convert a motor from scratch, including dismantling, sensor-mounting and wiring, and reassembly. If the motor

EXCESS LEAD LENGTHS are trimmed as the three separate sensors are put in place at A & C Electric. Then the interconnecting wires are dressed down tight to the windings with tie wraps.

is already disassembled, the retrofitting takes only about 15 minutes.

If these sensors eliminate motor burnouts, isn't A & C worried about a falling off of their rewind business? Not at all, they say. After all, there are about two million motors in the Detroit area, and they fail for many reasons other than burned-out windings. Also, the fact that users will tend to push sensor-equipped motors to the limit of their overload capability may mean that other component failures may show up in increasing numbers.

The company finds that, for the customer, the cost of retrofitting protection is such that the installation generally pays only where the motor is used in a critical process, is likely to be overloaded, and is on a machine where replacement is time-consuming and downtime is expensive. Interest outside the automotive industry is picking up slowly, but A & C believes

it may be quite some time before the units are really widely used throughout industry.

A & C has some suggestions for other shops thinking of offering this new service to customers.

1. Make sure you protect every phase in the motor.

2. Placement of the sensor ½ in. one way or another won't make any difference, but it should be placed approximately in the center (side to side) of the phase coil and not too close to the insulation between coils. (The insulation could affect the temperature the sensor "sees.")

3. Don't place the sensors at the end of the motor where the stator power leads are connected to the windings.

4. Make sure that the leads interconnecting the sensors are firmly tied down to avoid contact with other parts of the motor.

5. Make sure each sensor is in intimate contact with the windings. We make an effort to place the sensor in a crevice in the windings. While there is no evidence that this is essential to proper performance, all the motors we have retrofitted appear to be working well. A crevice provides more surface contact area with the sensor tip, which should enable the system to more closely track the actual winding temperature.

6. Use only the epoxy material for securing the sensors in place that is provided by the manufacturer of the retrofit kit. This epoxy has a very high heat-transfer coefficient. In one case when we were pushed to get a motor out the door, we didn't have the right epoxy, so we used a standard formulation we had around the shop. When we tested the motor, the sensor system did not work efficiently and we had to remove the sensors and wait for the proper epoxy.

7. Make sure you use the sensor specified for the motor temperature rating. One customer was so anxious for his motor that he gave us the go-ahead to use a sensor for a Class B motor (which was all we had on hand at the moment) on his Class F motor. The Class B sensor is calibrated to kick out at 135F. We got the motor right back from the customer because the sensor kept doing its job, and of course the Class F motor wasn't working at anywhere near its burnout point. You can't fool these sensors.

Know insulation systems for longer motor life

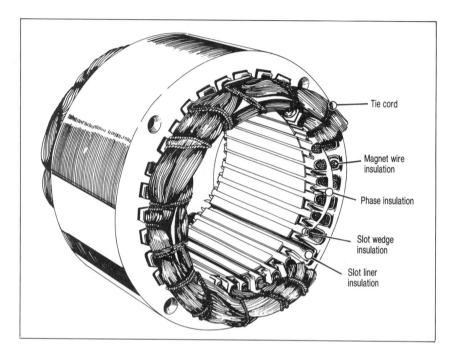

Tie cord

Magnet wire insulation

Phase insulation

Slot wedge insulation

Slot liner insulation

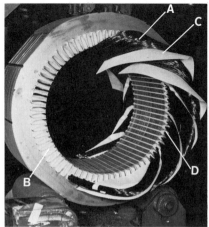

INSULATION SYSTEM for a typical random-wound motor is shown by photo and corresponding sketch, which is labeled with components of the system.

By WILLIAM I. KOLOVRAT
and VELLO NOLVAK

INSULATION is the heart of the electric motor. It is a major element that has the greatest effect on motor life and its maintenance history. When insulation fails, the motor fails. Insulation is the one component that undergoes more design changes than any other part of the motor, especially today when equipment sizes seem to be shrinking and motors are subjected to higher operating temperatures.

In addition, the reduction of energy costs, improved productivity, and reliability of operation have become of paramount importance to today's facilities managers. These electrical people, for whom cost and continuity of applied power are particularly important, are now taking a closer look at the key element in the performance

WILLIAM KOLOVRAT is Materials and Process Engineer, Large Motor Div., and **VELLO NOLVAK** is Engineering Manager, Insulating Materials Div., Westinghouse Electric Corp.

and dependability of the motors and generators—the insulation system. Therefore, it has become vital that chief electricians, plant electrical engineers, and other facilities electrical people obtain a good knowledge of motor insulation to aid in the establishment of an effective maintenance program and help make the proper insulation specification for a particular application.

The primary functions of a motor's insulation are to separate the various electrical components from each other and to protect itself and the components from such environmental contamination factors as dirt, chemicals, and other destructive forces—including heat and vibration.

The type of insulation system required depends upon not only the environment in which it operates, but also the characteristics of the insulating materials. The following checklist of causes and effects will help determine the insulation system needed.

● High ambient temperature (temperature of air surrounding motor, higher than 40°C or 104°F) causes deteriora-

tion and accelerated aging. It can also reduce viscosity of grease in bearings, thus reducing the value of lubrication.

● Corrosive agents (chemicals) eat away at insulation and exposed metal parts such as air-gap surfaces, bearings and shaft, eventually causing malfunction.

● Abrasive agents (sand, filings, dust) erode protective coatings as well as exposed surfaces, scoring critical bearings and wearing commutators and rings.

● Blanketing agents (dust, dirt, snow) can coat critical motor components, clogging air vents and passages. Heat dissipation is thereby reduced.

● Moisture (high humidity, rain) causes shorting of windings as well as rusting of metal parts.

● Mechanical abuse (shock, vibration) causes mounting feet, frames, shafts and brackets to break. It also causes chattering brushes, fatigued terminal connections, and damaged bearings.

● High-altitude operation (more than 3300 ft above sea level) causes excessive heat. Thinner air has less cooling ability.

Insulation system components

Major elements of a typical insulation system for an AC random-wound motor are shown in the accompanying sketch. The two most significant components of the system are the turn-to-turn or magnet-wire insulation (A) and the phase-to-ground or slot-liner insulation (B). The magnet-wire insulation separates individual wires in each coil. In smaller motors this insulation is usually enamel; in larger, form-wound motors, the coils are taped.

The slot liner insulates between the windings as a whole and the "ground," or metal parts, of the motor. This can be a composite layered sheet material of polyester mat/polyester film/polyester mat, which provides both dielectric and mechanical protection.

Phase-to-phase insulation (C) separates adjacent coils in the different phase groups. This is a separate flexible sheet on smaller motors but is not required on larger form-wound motors, where tape performs this function.

The tie cord (E) secures the coil winding end turns.

The wound stator can be dipped or vacuum-impregnated and then baked with an electrical varnish to provide adhesion, heal any wire damage from winding and handling, provide moisture and chemical resistance, and generally upgrade the insulation system's overall performance.

At about 500 hp, winding construction of AC motors changes from round-wire, random-wound coils to rectangular-wire, form-wound coils. Like the conductors in the random-wound motor, the rectangular copper conductors are coated with enamel. They are then insulated with glass, aramid fiber paper, or mica paper.

The slot section of the coil is wrapped with mica paper, and the end turns, or diamond sections, of the coil are insulated with mica tape. These sections are overwrapped with armoring glass or polyester glass tape.

In the stator of a large AC motor, varnished glass may be used as the slot cell for added protection during winding. The form-wound coils are then placed into the stator. Wedges may be of glass polyester, canvas phenolic, or other high-pressure laminate. A polyester felt material is used for the braces between the diamond sections of the coil. These braces make the end turns rigid to withstand the mechanical forces of start-up and protect against short-circuit conditions.

The completely wound, wedged and braced stator may be varnish-dipped or vacuum-pressure impregnated.

Temperature ratings

Motor insulation is rated sequentially by the thermal class letters A, B, F and H, each letter directly corresponding to a temperature index expressed in degrees centigrade and indicating maximum operating temperature.

Thermal class	Thermal index
A	105°C
B	130°C
F	155°C
H	180°C

For example, a Class B motor has a maximum allowable operating temperature, or hot spot, of 130°C.

This hot-spot temperature, which is the temperature at the center of the motor coil, is the sum of all the factors that produce heat in the motor winding. These include ambient temperature, motor rise temperature by resistance, and the hot-spot allowance (see accompanying table).

Some of the various insulating materials and their applications found in each classification are:

• *Class A* — Rag paper or polyester film. This class is used for fractional-horsepower applications: electric drills, vacuum cleaners, small appliances.

• *Class B* — Polyester mat/film/mat — a 3-ply composite can be saturated with polyester or epoxy resins. Considered the NEMA standard, most off-the-shelf motors have this insulation for HVAC application (refrigeration and air conditioning) and for fractional-horsepower motors.

• *Class F* — Polyester composite and aramid fiber paper or laminates of aramid paper. Many more motors today are using this insulation class to prolong operating life. Used for industrial applications (process lines, pumps, compressors and fans).

• *Class H* — Composites such as aramid fiber paper and polyester film, polyimide film, or combinations of the foregoing materials. High-temperature, high-reliability applications for transportation industry (subway, people movers), heavy construction products and for high-ambient, high-altitude use.

Temperature is vital

Although mechanical and environmental factors affect the life of a motor insulation system, the primary limiting force is heat. How long any given insulating material will last depends on the degree and duration of heat to which it is exposed.

Electrical insulation is assumed to have a life expectancy of approximately 20 years if the motor is operated within specified (nameplate) limits. To determine how long a motor insulation system will last in a particular application, the industry rule of thumb is: for every 10°C increase in insulation temperature, its life is halved; and for every 10°C decrease in temperature, its life is doubled.

Of course, other factors affect motor insulation life. For example, during motor start-up, as much as six times the normal current can be applied to the motor, producing more heat. While this is only a momentary surge through its windings and has little effect, if your motor is required to start many times each day or if frequent starts with an unusually heavy load are necessary, it would be wise to consider a better insulation system.

The motor user can be assured the stock motor he buys from a qualified salesperson or distributor is correct for the application specified. For example, for a motor operating in a typical plant where the air is relatively clean and dry and motor abuse is absent, a Class B or F insulation system in a standard off-the-shelf motor should be adequate. On the other hand, if the motor is required to work in the hot, dirty atmosphere of a mine, then clearly a more sophisticated system is needed.

Hot-spot temperature and allowances (°C)

Insulation class letter	Temperature	Allowances		
		Open	TEFC	TENV
A	105	5	5	0
B	130	10	10	5
F	155	10	10	5
H	180	15	15	5

INSULATION CLASS depends on type of materials used and correlates to a "hot-spot" temperature rating. Standard hot-spot allowances have been established for the insulation classes and basic motor enclosures. The hot-spot allowance is the difference between the rated temperature at the center of the coil and its surface temperature.

Insulation-resistance tests assure motor reliability

Insulation-resistance testing plays a key role in a well-planned motor maintenance program for a large New York City skyscraper.

THE ELECTRICAL preventive maintenance program at the world headquarters of the Equitable Life Assurance Society of the United States is vital to continuous operation of electrical systems serving thousands of people in this 42-story skyscraper.

An essential part of their program is the maintenance of hundreds of motors of various sizes and types. All critical motors, as well as those that are large, costly, and hard to replace, receive regular maintenance. Insulation-resistance testing plays an important role in keeping these motors running.

For example, the air-conditioning systems for the huge building utilize several large motors ranging in size from 100 to 500 hp. A failure of one of these motors during a hot spell would result not only in personnel discomfort, but also in overloaded equipment, which would then become more vulnerable to failure. To prevent this, any suspect motors are checked out during winter months.

In addition, several of the motors are no longer manufactured. This means that temporary replacement is next to impossible, and permanent replacement would involve extensive changes to mounting arrangements, coupling to driven equipment, and changes in control equipment. This, of course, would result in an extended downtime and expensive replacement costs.

As a result, this equipment receives careful attention within the framework of the building electrical preventive maintenance program. Years of experience, coupled with accurate record-keeping, has helped John Pensenhofer, chief electrician, to determine the most effective frequency of inspection and tests. For most motors as well as other important electrical equipment, inspections and tests are performed on an annual basis.

According to Pensenhofer, dirt, heat, moisture and vibration are arch enemies of electrical equipment, and can do a great deal of damage to insulation, bearings, contacts and most moving parts. Therefore, the heart of their motor maintenance program is a visual inspection backed by insulation-resistance tests.

Insulation test techniques

Insulation-resistance tests are carried out using a battery-powered instrument that permits reliable, non-destructive testing and evaluation of electrical equipment. The instrument is easy to use because it is pushbutton-operated and provides for continuous testing without hand-cranking.

The electrical crew uses two basic insulation-resistance tests: a *short-time test*, in which the resistance of the motor windings is checked quickly, and a *comparative short-time test*, which is done annually and which provides readings that are compared and evaluated each year.

The short-time test is often performed when it is desirable to obtain a quick evaluation of the condition of a motor. Usually three readings are taken—one from each phase of the motor to ground. If all three readings are above acceptable minimum values, the motor is considered operable for a preselected period of time (usualy six months to a year). Presently, the acceptable industry practice permits *one megohm* as the absolute minimum value of insulation resistance for a 460-volt motor. Minimum insulation resistance normally should be somewhat higher—in the range of 20 to 50 megohms. However, acceptable values will vary in accordance with many factors, such as voltage rating of the insulation, type of insulation, altitude, environment, and hp rating of the motor. Of particular significance are the effects of temperature, humidity and cleanliness of the area.

The comparative short-time test provides a highly reliable evaluation of the condition of the motor insulation. This technique is essentially the same as the short-time method except that readings are taken for a longer period of time; the instrument remains connected for approximately 10 minutes each time a reading is obtained. Readings usually vary in accordance with the length of time the instrument is connected. However, at the end of 10 minutes (which is an arbitrary length of time; 15 or 20 min could be used) the readings obtained usually reach a steady-state condition. On good, dry insulation, readings will normally climb slowly for about 10 min. If the insulation is damp or dirty, a steady state is usually reached quite soon after the test is started. All readings

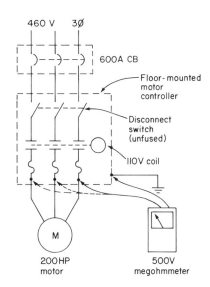

CIRCUIT DIAGRAM for a typical insulation-resistance test on a large motor serving an air-conditioning system.

SAFETY TECHNIQUES include turning off of supply circuit breaker, as electrician in background is doing, and checking for presence of voltage at motor controller using battery-powered insulation-resistance tester set to operate on its "voltage" scale. Note that only one hand is needed to obtain voltage reading.

are taken annually and repeated for a period of years. Values thus obtained are charted, compared and interpreted, enabling a more accurate evaluation of the condition of the motor insulation. See accompanying illustrations.

Test procedure. At Equitable a team of two electricians carry out the test. Test procedures are directed by Jake Provost, senior electrician, who is often assisted by Anthony Faranda, plant electrician. First, Provost reviews all data on the motor to be checked. These records not only give him an idea of what to expect; they also warn of past problems and help him to prepare for safe, effective testing.

The accompanying photos and circuit diagram typify the testing of a 200-hp, 460-volt induction motor. The motor controller is a floor-mounted unit furnished with an unfused disconnect switch and full-voltage contactors. Branch-circuit protection is provided by a 600-amp, molded-case circuit breaker installed in a nearby panelboard.

Safety is a primary concern. Before connecting the instrument, the senior electrician makes certain that all equipment scheduled for the test is completely disconnected from all power sources. In this case, initial steps include the opening of both the circuit breaker and the disconnect switch.

Next, Provost uses a tester to make sure voltage is not present at the input terminals of the motor controller. Because the tester can also serve as a voltmeter, he can set the instrument to

read volts (see photo). Note that one lead of the tester is provided with a standard probe and the other with an alligator clip. The lead with the clip can be attached to the appropriate terminal or ground. Thus, the electrician

makes contact with the other insulated probe with only one hand in proximity to possibly energized equipment. He must be *certain* that voltage is *not* present before starting the insulation-resistance test.

ELECTRICIAN inspects contacts of motor controller (note arc chutes on floor) and checks for general condition of the equipment. All observations are recorded for future reference.

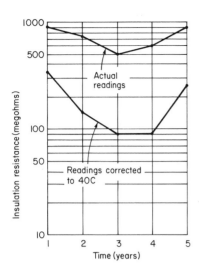

INSULATION-RESISTANCE READINGS are plotted against time (years). Trend of readings is more important than any individual reading. A continuing downward trend, or a significant drop in value, indicates failing insulation.

After all safety checks are completed, the insulation-resistance readings can be taken at a point closest to the motor itself. In this instance, the most convenient point was at the back of the motor controller, where conductors from the motor come to the "load"

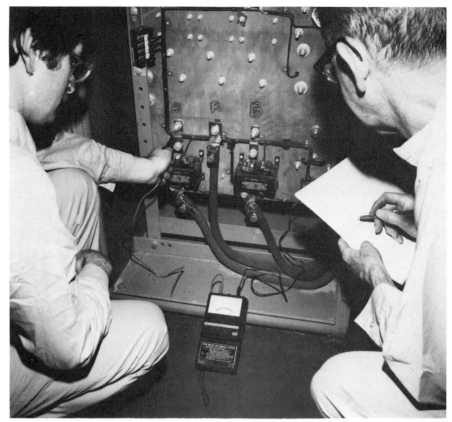

INSULATION-RESISTANCE READINGS on 200-hp motor are taken at convenient location at back of motor controller. One lead is clipped to frame (ground) of enclosure. Electrician at left, Anthony Faranda, touches probe to each of the three conductors coming from motor. Probe is in contact with each terminal for two minutes. Senior electrician Jake Provost, at right, records readings.

TEMPERATURE in area of motor can have a significant effect on the insulation-resistance readings. Jake Provost checks temperature to permit valid comparison of readings.

Correcting insulation resistance for ambient temperature

CONVERSION CHART

Insulation resistance coefficient K_t (y-axis)

100
50

10
5

1.0
0.5

0.1
0.05

-10 0   50   100

Winding temperature t (deg C)

READINGS

Insulation resistance: 900 megohms
Temperature: 22C
Relative humidity: 30%

CORRECTION TO 40C

$$R_c = K_t \times R_t$$

where

R_c = insulation resistance (megohms) corrected to 40C

R_t = measured insulation resistance (megohms) at temperature t

K_t = insulation resistance temperature coefficient at temperature t (from conversion chart)

CALCULATION

$$R_c = 0.3 \times 900 = \textbf{270 megohms.}$$

RELATIVE HUMIDITY can also have an effect on insulaton-resistance readings. Electricians use a sling psychrometer and slide-rule type computator to obtain relative humidity readings.

terminals (see photo on opposite page).

Preceding the test, an "infinity" reading is checked with the instrument leads separated, followed by a "zero" reading with the leads shorted. This provides a check on the instrument accuracy.

Next, the alligator lead is clipped to ground (frame of the controller), and a reading is taken at each motor lead terminal. The tester probe is placed in contact with each terminal for 10 minutes, after which the reading is recorded. In this case, the reading obtained on all three terminals was 900 megohms. This high reading is a strong indication that the motor insulation is in excellent condition. Ambient temperature readings and relative humidity readings are also taken and recorded, together with the conditions in the area—wet location, excessive dust or corrosion, presence of any chemical fumes, whether the motor had been running or had been at rest prior to the test. These factors have a significant effect on insulaton-resistance readings and aid in interpreting the readings.

The insulation-resistance readings are next corrected for ambient temperature variations to a standard 40C. A temperature correction graph and sample calculations are provided in an accompanying illustration. (The correction formula, graph, and other helpful data are available in *IEEE Recommended Practice for Testing Insulation Resistance of Rotating Machinery*—ANSI/IEEE Std 43-1974.)

All insulation-resistance readings are recorded annually on a time vs insulation-resistance graph. (Where three separate readings are taken, the lowest value is used.) Both the actual value and the value as corrected to 40C are entered on the graph. The readings are then plotted as shown. Insulation normally will show a slight decrease in resistance as it ages. The readings plotted show a relatively fast decrease in the first three years, with a return to normal at the fifth year. There is no question that this insulation is good; variations in the readings may be caused by extensive clean-up of surroundings, conditions when readings were taken, or a variation in test method. Had one of the readings dropped sharply to say 2 megohms, it would have indicated a possible—even probable—imminent insulation failure. In this event, additional investigation would be undertaken.

Electronic protective relays assure motor reliability

By **FRED SHIRZADI**

AN AC INDUCTION motor is a dependable device that provides long and trouble-free performance; however, it must be protected from numerous types of malfunctions that could be caused by outside sources. Such problems include thermal overload, single-phasing, ground faults, underload, overvoltage, stall conditions, and incorrect phase sequence or reverse-phase operation. Protection against such faults will prevent premature motor failures, eliminate expensive downtime, reduce required maintenance, and generally boost productivity of the process or driven machine.

Why do motors fail? Recently, the Electrical Research Association (ERA) in Leatherhead, England, studied instances of damage to more than 9000 motors in England, Finland and the United States. The study indicates the relative importance of the individual causes of damage. This investigation revealed a typical fault rate of 2.5% per year and demonstrates major causes of failure, as shown in the accompanying table.

Approximately 25% of the motors examined were rated 50 hp and over. Repair costs for this group, however, represent 80% of total repair costs. This emphasizes that larger-size motors need good protection, careful system design, and proper maintenance of the motor and associated drive. To give an idea of the repair costs involved, it is estimated that in England alone annual costs run to about $40 million.

Note in the table that the thermal overloading (caused by continuous overloading and also by stalling) is the major contributor to motor failures— 30% of the total. This overloading usually results in uniformly burnt-out windings. Of particular interest is the fact that these faults occurred in spite

FRED SHIRZADI is chief engineer of Sprecher & Schuh, Inc., Port Chester, NY, where his responsibilities include quality control and research and development.

of the use of normal protection schemes. Note also that an average of 14% of all defective motors failed because of single-phasing. In 19% of the failures, winding or bearing damage was due to contamination, oil, humidity, etc. And 13% of the motor failures were caused by bearing damage; 10% were due to aging of the terminals or windings; 5% were due to rotor problems.

What must modern motor protection do? Many motor failures could be prevented by proper preventive action. That is to say, bearing failures are frequently caused by excessively tight pulleys, poor alignment, and similar mechanical faults; contamination failures could be prevented by better encapsulation and maintenance; and old-age failures could be prevented by timely replacement of worn-out components. However, overload and single-phasing, which represent 44% of all failures in the study, must be taken care of by properly selected motor protection devices and systems.

A modern motor protection system must provide advance warning of a fault to an operator (time permitting) or disconnect the load before damage occurs. In some cases, however, disconnection before fault occurrence may be impossible. Then the protection system must detect the fault condition as quickly as possible and minimize any damage during fault operation of the system.

In addition to providing maximum equipment protection, a well-designed system must maximize the up-time of the overall system. In other words, ideal protection trips the motor only if there is an impending hazard and does *not* cause unwanted (and often very expensive) downtime. Ideal protection can be further defined as such that results in optimum operation and control of the installation. An example of this is a motor-start prevention system that blocks the starting attempt of a heavy-duty motor until there is sufficient thermal reserve in the motor to permit a start without overheating. A fruitless (and possibly damaging) starting attempt thus might be avoided.

Causes of motor malfunction	Percent
Overload	30
Single-phasing	14
Contaminants	19
Old age	10⁻
Bearing failures	13
Rotor failures	5
Miscellaneous	9

Base: 9000 failure events (study by Electrical Research Assn., Leatherhead, England)

MAJOR CAUSES OF MOTOR PROBLEMS are listed. Thermal overloads and single-phasing make up 44% of malfunction causes. Approximately 80% of these problems occurred in motors rated 50 hp or larger.

Another example is a stalled-motor protection system which, with a serious fault in the drive, does not wait until the thermal protection system responds but disconnects immediately, thus avoiding an unnecessary motor temperature rise (and the associated long cooling period prior to a safe restart of the motor). This essentially means that, for optimum effectiveness, a motor-protection system must be fully integrated, combining all required protection functions in an integrated package working toward a common goal—to maintain proper motor operation effectively, not just merely to prevent damage to a motor.

How much heat can a motor take? Since thermal overloading is among the leading causes of motor failures, it is well to examine just how hot a motor can get before damage occurs. As the ERA study showed, the most thermal-critical parts of a motor are the stator windings. Depending on the class of insulation, the insulation material is selected for certain continuous operating temperatures and is a deciding factor in the life of a motor. In case of class B insulation, the average winding temperature is 120°C. Under fault conditions, disconnection must take place as soon as the winding temperature reaches 165°C maximum. For short periods of time, however (as, for instance, during a heavy-duty start), higher temperatures can be permitted without any significant effect upon the motor life.

Motor specifications listed on the

nameplate are always referred to at an ambient temperature of 40°C. Unless the manufacturer indicates otherwise, this means that the motor will deliver its full rated output at this temperature without overheating. At lower ambient temperatures the motor can handle greater loads; at higher ambient temperatures its loading must be reduced.

The operating temperature of the motor windings determines its insulation and life. Assuming continuous operation of a motor at its rated temperature of, say, 120°C and Class B insulation, a motor would have a life of 10 years and the aging process would be linear. If however, the same motor were subjected to 130°C operating temperatures, *its life would be cut in half*, i.e., to 5 years. With the operating temperature of 140°C, its life would be cut in half again; i.e., down to 2.5 years, and so on. (See accompanying thermal curve.)

The above shows the importance of correctly setting the thermal protection system to keep the motor temperature as low as possible and thus get the most life out of the motor. Simple bimetal relays may permit a motor to run at 140 or 150°C, greatly reducing its life.

Which thermal protection is best? Essentially, there are two basic thermal-protection techniques—one relies on some means of measuring the winding temperature directly, and the other evaluates the motor winding temperature on the basis of current drawn by the motor.

On the surface, the direct winding-temperature measurement method seems like the preferred way. Indeed, what could appear more reliable than a direct measurement? By placing a suitable temperature sensor within a winding (either solid-state or electromechanical), one can obtain a continuous temperature reading. Using suitable circuitry, then, it is a simple matter to develop a protection system that would be fast, simple, and reliable.

However, the most serious objection to direct-measurement devices is the fact that they can only measure the winding temperature locally. The sensor always measures the temperature at the point where it is located and is incapable of detecting temperature differences in other parts of the motor. Also, it is incapable of detecting an overload condition that could result in a motor burnout. Some other common motor faults that such direct-measuring sensors cannot guard against include rapidly occurring single-phasing, a short-circuit, and ground leakage. And it is only under certain conditions that a false restart attempt can be prevented.

In addition to these operational shortcomings of direct temperature sensing, there are several practical

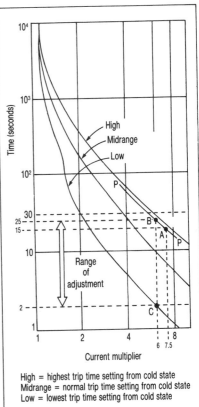

TIME/CURRENT CHARACTERISTICS permit accurate setting of trip time in accordance with application. As an example of their use, assume a motor's starting current is 7.5 times its full-load current and its permissible stall time is 15 sec. Move up from 7.5 on the horizontal scale to point A, where this line intersects the horizontal 15-sec line. Draw line segment P-P through point A and parallel to the "high" characteristic curve. Since a starting-current value of six times the full-load current is empirically accepted, draw a vertical line from 6 on the horizontal scale to point B on line segment PP. Move left to the vertical scale and read 25 sec as the maximum acceptable OL relay tripping setting. Point C represents the minimum setting, and the range of adjustment is from 2 to 25 seconds.

objections as well. For instance, the cost of fitting thermistors and the associated tripping device can be high. Also, a request for in-winding sensors tends to increase motor delivery time.

Another consideration is the ability to perform a functional check. Trip circuits, especially those associated with thermistors, are arranged so that protection actuates a trip whenever the lines to the thermistors are interrupted. Checking the response temperature and the quality of the thermal coupling at the winding, however, is

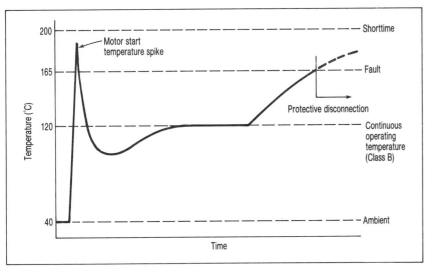

THERMAL CURVE shows that high temperatures reached for a short period during starting will not trip out the motor and is not harmful to Class B insulation. However, when temperature climbs to 165°C for any continuous length of time, the relay should trip, taking the motor off the line. The adjustable trip time is selected in accordance with requirements of the motor application.

virtually impossible. Yet, a thermistor installed in a motor winding is often subjected to severe mechanical stresses, and removal of a damaged thermistor from motor windings is not a straightforward matter.

A motor's operating conditions are quite accurately described by the amount of current it draws. With correct interpretation, the input current can serve as a measuring stick for determining when to shut off a motor. Indeed, the current drawn by a motor at any given time uniquely defines the amount of electrical power that the motor is consuming. This, of course, directly relates to the temperature rise in the motor windings. In addition, the electric current is quite easy to measure in comparison with several other physical variables that could also be used to monitor motor performance.

Therefore, it appears that a protection system that judges the motor condition strictly on the basis of motor current and is entirely external to the motor provides optimum protection as well as functional flexibility.

Determining the shutdown criteria. During motor startup and under high overloads, the major power losses occur in the windings (stator and rotor), where the thermal capacities determine the permissible load and the permissible starting time. By adjusting the tripping time of the OL protective relay, its characteristics can be tailored to the individual motor. A suitable starting point for performing this adjustment is the permissible stalling time of the cold motor in conjunction with the corresponding current.

Typical time/current characteristics used to adjust tripping time in a modern electronic motor protection unit are shown by curves in the accompanying sketch. The caption shows how these curves are used.

This procedure will also result in adequate protection even under the following applications and conditions:
1. Heavy starting current (fans, centrifuges, elevators, etc.)
2. Explosion-proof motors
3. Submerged motors
4. Hermetically sealed refrigeration compressors.

The adjustment of the tripping time permits the thermal capacity of the motor to be used in the most cost-effective manner while reducing the possibility of nuisance trips. If a plant with motors running at their thermal limits is shut down, the motors must be given sufficient time to cool off before restarting them. An attempt to restart a hot motor will fail, since the protective system will trip and disconnect the motor.

Other motor protection functions. A modern electronic motor protection system can sense and display numerous types of motor faults by monitoring the motor current. Fault displays make troubleshooting quick and easy by pinpointing the nature of each fault.

To begin with, consider a fault associated with phase loss (popularly known as "single-phasing"). Even though the danger to large motors in this case is well known, the possibility of its occurrence is quite real—a blown fuse on one phase will result in single-phasing. This condition, however, can be easily detected with an electronic motor protection system independent of motor loading by simply monitoring each phase. Thus, a motor can be shut down as soon as single-phasing is spotted.

Next there is ground-fault detection. Most insulation faults in motors result in a leakage to the grounded parts of a motor. As a rule, it is practically impossible to prevent the occurrence of these faults. However, it is possible to detect them quickly and limit the damage to a minimum.

In the insulated-neutral system (with high-impedance grounding) or the resonant-grounded system (with ground-fault neutralizers), only relatively small leakage currents occur, which usually allow the motor to remain in operation for a short time. While central neutral-point monitoring reports the occurrence of a ground fault in general, the motor branch circuit involved can be indicated separately, permitting required repairs.

In the grounded-neutral system, ground-fault currents can build up rapidly to short-circuit magnitude, in which case the motor must be shut down quickly to limit the damage. This is achieved by an electronic protection system.

With the large, heavy-duty motors used in industrial plants, occasional mechanical jamming or stalling leads to extremely heavy currents. Under these conditions, it is best to disconnect the motor as soon as possible. This eliminates unnecessary mechanical and thermal loads on the motor and damage to the associated power-transmission train.

To prevent nuisance trips, however, the system must differentiate between a current surge caused by a startup and a surge due to mechanical jamming or stalling while in operation. This is accomplished by a logic circuit that enables protection only after the motor has been started and is running. This protection usually disconnects the motor within a fraction of a second after jamming occurs.

Another type of fault is an underload condition, when the motor current falls below some minimum value. This might very well indicate a problem in the overall system—a broken gear in the power train (effectively removing the mechanical loading from the motor), or clogged filters on fans. In the event of such a fault, the underload protection responds after a suitable time lag. The trip current or signal for the underload protection can be adjusted over a relatively wide range to accommodate a variety of motors and conditions.

Finally, there is always the possibility of a motor running in the reverse direction (with the wrong phase sequence). This can result in significant damage to the motor and possibly to the associated power-transmission train. Furthermore, improper phase sequencing may also cause accidents, injuring plant operators.

As in single-phasing, this condition can be sensed independent of motor loading; the protective system simply monitors the phase sequence and responds if it is wrong.

Making motors last longer

By **JOSEPH SMITH**

MOTORS LAST LONGER and run better when all mechanical components fit together properly. According to Joseph Smith, general manager, R. Scheinert & Son, Philadelphia, PA, most motor failures are the result of mechanical malfunction. Most mechanical malfunctions relate to the motor bearings, bearing housings, shaft and end bells. Because of this, the firm provides special attention to these components and suggests that plant operating and maintenance personnel inspect and maintain these parts on a regular basis.

When a motor comes in for repair or rebuild, Scheinert technicians carefully inspect bearings and housings for dirt or contamination, damage, dents, scratches, out-of-round balls or races, or distortion of any kind. They also check the shaft and end bells for wear

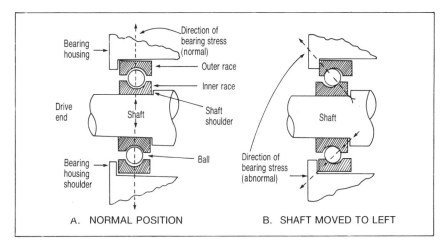

A. NORMAL POSITION B. SHAFT MOVED TO LEFT

CROSS-SECTION OF BALL BEARING is shown by sketch. In A, the bearing is in a normal position with room for end play between outer race of ball bearing and shoulder of bearing housing. Normal stress is at right angles to the shaft, resulting in normal bearing operation. In B, the shaft has moved to the left, but the bearing housing and motor enclosure remain stationary. Note that the inner race moved with the shift and the outer race moved until it pressed against the bearing housing, where it is held in place. The balls have been displaced so that only minimum contact points occur. This causes extreme pressure at these points, and force is applied in an angular (abnormal) direction, resulting in excessive stress, friction and heat that causes early bearing failure and short motor life.

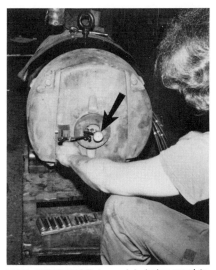

DIAL INDICATOR (arrow) is being used to check end play of shaft on a rebuilt motor in an electrical-mechanical service firm. Adequate end play (movement of the shaft/rotor in an axial direction) is vital to proper operation of the motor.

JOSEPH SMITH is general manager of R. Scheinert & Son, Inc. Electrical-Mechanical Service & Sales in Philadelphia, PA.

or damage. If the shaft is worn or damaged in any way, it receives flame-spray metallizing and is turned down back to original specifications.

After repair to the motor is completed—rewind or reconditioning—all components are checked for proper size and fit. Measurements must be accurate. Tolerances must permit all components to expand during operation of the motor and yet maintain proper "freewheeling" with no binding that could cause overheating and eventual failure.

Next, the motor is reassembled and rotation is checked for freedom. Also, as shown in the accompanying photo, end play of the rotor/shaft is carefully checked. The technician in the photo is measuring the amount of end play with a dial indicator mounted on the end bell. The amount of movement should match the manufacturer's specifications. Smith points out that the proper amount of end play is critical, because as the shaft expands during operation

of the motor, enough room must be available to avoid binding of any sort that would shorten bearing life. All motors will have some amount of end play in the range of $\frac{1}{32}$ in. to $\frac{1}{16}$ in. Some designs however, have a locked bearing at one end, which results in the rotor not moving in an axial direction. Therefore the opposing bearing must be able to move or "float" to provide the required freedom. This type should be checked only by experienced personnel.

Primary causes of excessive, damaging end play include worn bearings or bearing housing, a worn or bent shaft, or contamination that could cause binding, wear and then excessive end play—in that order.

Usually, excessive end play is hard to detect during normal operation and inspection unless it is extreme. If it is suspected, it is best to call in a motor-repair expert, because a thorough analysis of the problem should be performed.

Know bearing basics for longer motor life

By **RAYMOND A. GUYER, JR.**

ROLLING-TYPE BEARINGS are classified into two basic families—ball and roller. Each family includes a variety of bearing designs, depending on speed required, operating temperature, and the kind of working load to be encountered. Sleeve bearings, special bearings, and custom designs have many applications; however, rolling-type bearings are used extensively in electric motors.

Ball bearings are found in practically any type or size of motor. They offer low friction, can operate at high speed, and run effectively over a wide temperature range. Modern ball bearings are well designed for the job at hand, are made of increasingly better materials, and last longer. Fig. 1 shows the basic parts of a rolling-type bearing.

In the ball-bearing family, the most popular assembly is the single-row deep-groove bearing (Fig. 2). It is a radial-load bearing, but it can also handle considerable thrust loads in either axial direction. (A thrust load is a pushing force against the bearing parallel to the shaft; radial load is a pressing force at right angles to the shaft. Most bearings are expected to carry both types of loads to some degree.)

The *angular-contact ball bearing* supports a heavy thrust load in one direction, sometimes combined with a moderate radial load (Fig. 3). The standard type can be used when the bearing is mounted singly. Modified types are used in duplex mountings for all combinations of thrust and radial loads (see Fig. 4). The sides of the bearings are universally flush-ground, permitting them to be installed in any way needed to do the job required. The preloaded type is mounted to handle the same load conditions as modified bearings, but it has a built-in factory

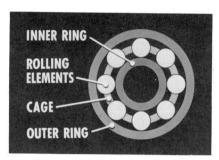

FIG. 1. Basic parts, rolling-type bearing

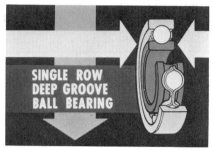

FIG. 2. Single-row, deep-groove bearing

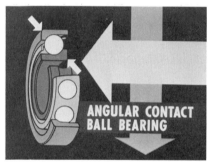

FIG. 3. Angular-contact ball bearing

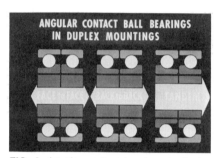

FIG. 4. Angular-contact bearings in duplex mountings

RAYMOND GUYER is supervisor of service engineering with the Technology Services of SKF Industries, Inc., King of Prussia, PA. He directs company bearing maintenance seminars held five times yearly.

preload. Preloaded bearings have an identification letter followed by a number that indicates the preload. Preloaded bearings should always be replaced with those having the same preload. Markings used by different manufacturers will vary.

Preloading refers to a bearing that has no axial or radial clearance. Radial clearance is the distance the inner and outer rings of a bearing can move radially in relation to each other when the bearing is unmounted. Axial clearance is the distance that the inner and outer race can move parallel to the shaft. There is virtually no internal movement in a mounted preloaded bearing. Some bearings have special radial

clearance. The higher the operating temperature, the greater the internal radial clearance should be. Since the bearing inner ring will expand more than the outer ring under high temperature use, a high radial clearance allows for heat expansion. To achieve maximum bearing life, bearings having special clearance designations should always be replaced by bearings with identical designations.

The *double-row deep-groove ball bearing* (Fig. 5) will sustain considerable radial loads because of the two rows of balls. It can handle moderate thrust load in either axial direction as well. Its capability is about 1.5 times that of a single-row bearing.

The *self-aligning ball bearing* (Fig. 6) has two rows of balls rolling on a spherical surface of the outer ring. This arrangement compensates for angular misalignment resulting from errors in mounting, shaft deflection, and distortion of the foundation. It is impossible for this bearing to exert any bending influence on the shaft, a most important consideration in applications requiring extreme accuracy at high speeds. Self-aligning ball bearings are used for radial loads and moderate thrust loads in either direction.

Ball bearings are available open or with seals and/or shields. These are designed to keep dirt and abrasives out of the bearing and keep lubricant inside. When a bearing is equipped with a seal, shield, or both, its outer dimensions do not change. It can fit within the same boundaries as an open or unprotected bearing. The shield alone is suitable for many applications. It keeps larger dirt particles out of the bearing raceways. Where dust and abrasives are present, a seal can be used. This seal is a stamped metal plate similar to a shield but with a rubber fabric washer attached to the inner surface. The washer lip rides on the seal groove of the inner ring. There are bearings that have a seal on one side and a shield on the other side, or only one seal or one shield. Some bearings have two shields or two seals.

Bearing numbers have a very important meaning for ball bearings. One manufacturer's system is typical of bearing identification methods. The first number indicates the manufacturer's series. The second gives the section height or outer diameter for a given size of bore. The last number, when multiplied by five, gives the size of the bore in millimeters. This is true for a 20- through 480-mm bore bearing. For example, a bearing is marked with the number "6210." The 6 indicates the bearing series (single-row, deep-groove ball); the 2 shows the diameter series; and 10 × 5 equals 50 mm, the bore size. More details on bearing numbering systems can be obtained from bearing manufacturers' catalogs.

Roller bearings are encountered in some large motors and in various portions of a drive train.

The spherical roller bearing (Fig. 7) uses two rows of barrel-shaped rollers. They allow the bearing to compensate for slight misalignment of the shaft. Although it is a very rugged radial

FIG. 5. Double-row, deep-groove ball bearing

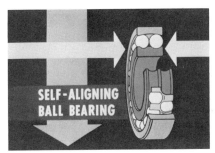

FIG. 6. Self-aligning ball bearing

FIG. 7. Spherical roller bearing

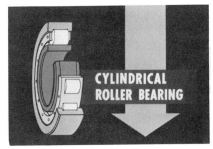

FIG. 8. Cylindrical roller bearing

FIG. 9. Tapered roller bearing

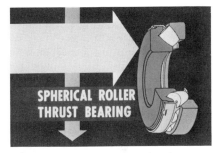

FIG. 10. Spherical roller thrust bearing

bearing, the spherical roller bearing can also carry considerable thrust loads in either direction.

The cylindrical roller bearing (Fig. 8) has a high radial load capacity in relation to its size. It is a low-friction bearing that is excellent for high-speed operation. A double-row cylindrical roller bearing is made with extra-precise tolerances.

The tapered roller bearing (Fig. 9) carries almost equal radial and thrust loads. It consists of an inner ring, called the cone, the rollers and their cage, and the outer ring, called the cup. This type of bearing should not be mounted on a shaft by itself, since the radial load induces a thrust load. To counterbalance the thrust load, a second tapered roller bearing is required.

Also available is the spherical roller thrust bearing (Fig. 10), which is specifically designed for extra-heavy thrust loads in one direction. It is inherently self-aligning, providing long life. Typical applications include on hollow-shaft vertical motors used in deep wells and for power-plant cooling water.

Bearings should always be replaced with their exact equivalents. Factors such as load, speed, operating temperature, environment, and life expectancy should be carefully checked to make sure replacement bearings are identical to the originals in all these respects. For any repeating bearing problems, a local distributor or the bearing manufacturer should be contacted for a thorough bearing review.

Proper bearing replacement assures reliability

By RAYMOND A. GUYER, JR.
SKF Industries, Inc.
King of Prussia, PA

CORRECT MOUNTING and dismounting of bearings, particularly roller/ball types, is vital to maximum bearing life and dependable operation. In recent years, ball bearings have been commonly used in integral-horsepower motors because they offer many application and operation advantages. However, many motor failures can be traced to a bad bearing. Further analysis reveals, in most instances, that bearings fail because of incorrect selection, inadequate maintenance, and improper replacement procedures.

When replacing a bearing, it is essential that maximum cleanliness be observed. A new bearing should not be removed from its original package until immediately before it is to be mounted. Mounting should, wherever possible, be carried out in a clean and dust-free room and not in the immediate vicinity of dust-producing machines. First, clean up all tools and devices that would be used when mounting the new bearing. Get rid of any linty cloth in the area. Do not handle the bearing any more than necessary. Fingerprints can become a starting point for rust. Be sure to inspect the associated components of the bearing arrangement carefully and remove all burrs. Clean the shafts and abutment shoulders. Check the bearing seatings with regard to diameter and accuracy of form. Also, inspect the seals and replace them if they are damaged or worn.

New bearings come thoroughly coated with slushing compound to keep out air, moisture, and rust. When used with synthetic hydrocarbon oils and greases, the slushing compound does not have to be removed, since it is compatible with such lubricants. However, when synthetic oils and greases with synthetic oils are used, the slushing compound must be removed. Bearings to be lubricated with polyurea grease may also require the slushing compound to be removed.

Bearings are wrapped in specially manufactured, heavy-duty, waterproof, polylaminate paper. Be careful not to drop bearings or handle them roughly and do not expose them to extreme temperature changes that might cause condensation to form.

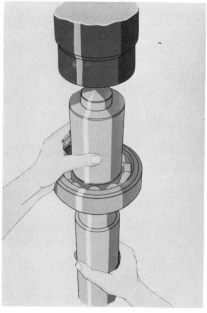

FIG. 1. Mounting press is used to install small bearings.

Heat should be used for mounting bearings larger than 4 in. in outside diameter. The most common heat-mounting routine involves the use of an electric hot plate that can maintain a steady temperature, a container full of clean oil, a thermometer, and a grating. The bearing is placed on the grating so that it does not touch the sides or bottom of the container. Direct contact would overheat the bearing, which would change the hardness of the metal. The oil is brought to a maximum temperature of 121°C (250°F), and the bearing is left in the oil from 15 min to 1 hr depending on size. When the bearing is thoroughly heated, slide it onto the shaft, put it in its proper location against the shoulder, and immediately lock it into position with a locknut—or hold it in place with a tube or pipe larger than the shaft. Unless the bearing is locked on properly, it may creep away from the shaft shoulder while it is cooling. If the bearing should get

caught while sliding up to the shoulder, it may seize on the shaft. If this happens, it is necessary to remove the bearing and try again. It should be noted that only open bearings (bearings with no seals or shields) can be heated in this fashion.

Another effective heating medium that gives automatic temperature control, eliminates fire hazards, and prevents rust is a 10 to 15% emulsion of soluble oil in water. The bearing is boiled for 15 min to 1 hr, depending on size, then removed and mounted as mentioned previously.

Other ways of heating a bearing are with a spiral heater and an induction heater. When using an induction heater, care must be taken to be sure the bearing is demagnetized. Never use a torch to heat a bearing, since hot spots can develop due to uneven heating and cause a softening of the bearing ring that would result in premature failure.

Mounting methods

Angular contact bearings are used where a lot of thrust and a heavy radial load are involved. For extra-heavy thrust loads in one direction, bearings are mounted in tandem. If thrust is exerted in two directions, they may be mounted face to face or back to back. Angular contact ball bearings can be mounted interchangeably back to back, face to face or in tandem, as shown by the cross-section drawings of Fig. 2, provided they are all of the same make, same size and number, and all are either modified or preloaded.

When a shaft is put in a vise in any bearing assembly or disassembly operation, protect the shaft from the jaws with sheets of brass or copper. It is very important to replace a bearing with another bearing of exactly the same size and type. Bearings and shafts are designed for each other, and changes cannot be made unless you are prepared to redesign the machine.

If the bearing fits too loosely on a shaft, it will creep or slip. This causes the bearing to overheat and also results in abrasive wear to the bore of the bearing and the surface of the shaft. However, if the press fit is too tight,

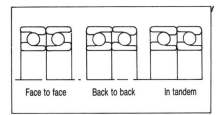

FIG. 2. Angular contact bearings are normally mounted in pairs.

the inner ring of the bearing will be stretched so much that there will be no room for the balls or rollers to revolve freely. Fit recommendations can be obtained by contacting the bearing manufacturer or local bearing distributor.

The use of a press is particularly suitable where small bearings are frequently mounted, as shown in Fig. 1. Place a sleeve between the bearing and the press. The end faces of the sleeve should be flat, parallel, and burr-free. It should be so designed that it abuts the ring which is to be mounted with an interference fit; otherwise there is a risk that the rolling elements and raceways may be damaged, and premature failure may occur.

To facilitate mounting and also to reduce the risk of damage, the bearing seating on the shaft and in the housing should be lightly smeared with a thin oil.

Small bearings may also be mounted by means of blows applied to a sleeve abutting one of the bearing rings, using an ordinary hammer. Hammers with soft metal heads should not be used because fragments of the metal may break off and enter the bearing. A suitable sleeve should have a welded end cover or preferably be made in one piece so that the hammer blows will be damped. If the sleeve is used frequently, it is advisable to provide it with a replaceable surface—for example, a threaded plug to which the hammer blows may be directed. When the bearing is to be mounted with an interference fit on the shaft, a sleeve that abuts the inner ring is chosen.

Preloaded bearings are designed to provide stiffness to the assembly. There is no internal radial clearance. Modified bearings, when mounted, have normal internal radial clearance. The difference between preloaded and modified bearings when both inner and outer rings are clamped together is shown in Fig. 3. Notice internal clearance in the modified bearing. It is

absolutely essential that the two outer rings of angular-contact bearings be mounted squarely within the housing. This can be accomplished by rotating the outer rings while the shaft locknut is being tightened. Several revolutions are necessary. Sometimes a mounting sleeve over the two outer rings aids in the proper alignment of the outer rings.

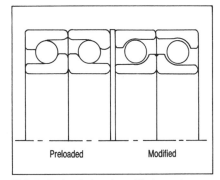

FIG. 3. Preloaded and modified angular contact bearings reveal differences in radial and axial clearance.

Spherical roller-thrust bearings must be mounted using the spring force on the bearing as recommended by the bearing manufacturer. If springs are not used and there is a possibility that the bearing can become unloaded during its operation, the rollers will skid, resulting in premature failure.

Single-row tapered roller bearings are always used in pairs and require an end-play adjustment (obtained with the aid of a dial indicator) for correct installation. The end-play adjustment is obtained by using shims for mounting the housing end plate, as shown in Fig. 4. The correct setting is where there is zero end play when the bearings are at operating temperature.

In mounting a separable cylindrical roller bearing where the roller and cage assembly and the outer ring are one unit, only the inner ring is first mounted on the shaft. Extreme care must be exercised in then mounting the outer ring and roller-cage assembly over the inner ring. If the outer ring and roller-cage assembly gets cocked, the outside diameter of the inner ring and rollers can become scored, leading to premature failure.

Dismount methods

In all instances, every effort should be made to remove a bearing without damaging it. This is, of course, particu-

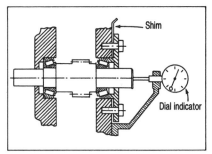

FIG. 4. Dial indicator and shims aid in obtaining proper end-play adjustment of single-row taper roller bearings.

larly important where the intention is to remount the bearing. The dismounting force should always be directed to that bearing ring with the interference fit.

A bearing which is to be reused should, for endurance life reasons, always be remounted in the same relative position as before. It is therefore advisable, before dismounting, to mark the position of the bearing—which side is uppermost and which side faces the front.

Small- and medium-size bearings may be dismounted using a conventional puller. If the bearing has been mounted with an interference fit on the shaft, the puller should engage the inner ring.

To avoid damage to the bearing seating, the puller must be accurately centered. The use of a self-centering puller eliminates the risk of damage, and dismounting is simpler and more rapid. Only in cases where it is impossible to engage the inner ring is it permitted to apply the puller to the outer ring. But, and this is important, the outer ring must be rotated during dismounting so that no part of the bearing is damaged by the dismounting force. This can be done by locking the screw and turning the puller continuously until the bearing comes free.

Dismounting the inner ring of cylindrical roller bearings can be easily done with an aluminum heating ring as shown in Fig. 5. The dismounting procedure is simple. First remove the outer ring with the roller and cage assembly. Coat the raceway of the inner ring with an oxidation-resistant oil. Heat the aluminum ring to a temperature of approximately 121°C (250°F), place it on the inner ring, and press the handles together. Use the tool to withdraw the inner ring as soon as it becomes loose. Remove the ring from the tool immediately. If the inner rings have different diameters and if dismounting is frequent, use of an induction heating tool is preferable, as shown in Fig. 6. Such heaters raise the temperature of the inner ring by inducing currents. The adjustable heater is suitable for various inner-ring diameters over 80 mm, depending on the manufacturer of the induction heater.

Heat the inner ring for 15 to 30 sec until it comes loose, and then withdraw it. The inner ring must not be heated to a temperature above 121°C (250°F). Switch off the current, remove the ring from the tool, and demagnetize it.

Used, open (not sealed or shielded)

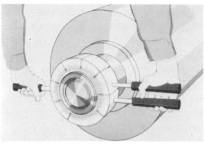

FIG. 5. Aluminum ring tool is heated to supply temperature of approximately 120°C to expand the inner ring for easy removal.

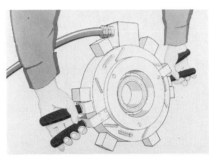

FIG. 6. Induction heating tool raises temperature of the bearing inner ring by inducing currents in the ring, expanding it for easy removal.

bearings, if heavily coated with oxidized grease, must be thoroughly cleaned before use. The bearings should be soaked in hot, light oil at 93° to 116°C (200° to 240°F), agitating the basket of bearings slowly through the oil. In extreme cases, boiling in emulsifiable cleaners diluted with water will usually soften the contaminating sludge. If the hot emulsion solutions are used, the bearings should be drained and spun individually until the water has completely evaporated and then adequately protected.

Prior to the startup of any equipment, the bearings should be lubricated in accordance with the bearing manufacturer's recommendation.

Service firm specializes in rebuilding and repair of explosionproof motors

FAN-GUARD PROBE is used to make sure that openings to the fan are inaccessible. Tests are being made here on two rebuilt 1½-hp, 440/220-volt explosionproof-motors.

GOODWIN Pray Co. Inc., Linden, N.J., affiliated with J. R. Longo & Sons, Inc., was the first service firm in the state to receive recognition under the new Underwriters Laboratories' program for listing rebuilt motors and generators for hazardous locations.

Until recently, an explosionproof motor requiring repair or rebuilding had to be sent back to the original manufacturer to be covered by the Follow-Up Service of Underwriters Laboratories, Inc. In other words, repair by anyone other than the original manufacturer would void the UL listing. Such seemingly inconsequential things as disturbing a motor's lead wire seal, covering metal joint surfaces with varnish, increasing clearances at shaft openings, placing shims between surfaces of assembled joints, or marring machined metal surfaces can adversely affect the explosionproof features of the motor, even though operation appears perfectly normal.

Because of many questions arising about the listing status of rebuilt explosionproof motors and to decrease the time and expense involved in shipping motors, particularly large ones, to the original manufacturer for repair, UL developed a program for listing motor and apparatus shops capable of rebuilding explosionproof equipment, together with a procedure for listing such rebuilt equipment. The firms recognized by UL for such work are listed in the UL *Hazardous Location Equipment Directory* (Red Book), issued annually, under the heading "Motors and Generators, Rebuilt."

Goodwin Pray engineers point out that a motor submitted to them for repair *must* bear an original UL label listing it as a classified-location motor before they can begin work, since the original strength and ability of the listed motor to contain explosions (barring undue corrosion or stress cracks) is assumed to still exist.

After a hazardous-area motor has been repaired and tested, it receives a special label stating that the machine has been rebuilt and is listed for use in a particular classified location.

The repair techniques, skill, and facilities of Goodwin Pray, as illustrated by the accompanying photos, provide an indication of the capabilities expected of service firms doing such work.

GOODWIN PRAY technician (above) uses 0-to-18-in. calipers to measure inside diameter of 25-hp explosionproof motor stator. Work requires a variety of high-accuracy micrometers, verniers, and calipers.

MICROMETERS used (right) range in size from 18 to 40 in. Machinist shown is measuring outside diameter of motor end bell. Vernier calipers and inside measuring rods are in foreground.

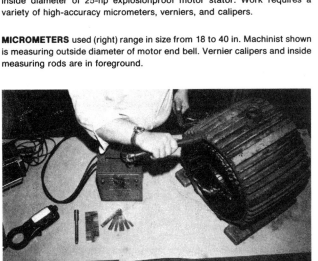

HIGH-POTENTIAL of 2500 volts ac is applied between windings and frame of 25-hp, 440/220-volt, TEFC, explosionproof motor (above) as part of a dielectric withstand-voltage test. Also performed after repair are continuity, ground, and surge-compression tests, as well as dynamic balancing. Instruments on bench here include (counterclockwise from left) an insulation-resistance tester, clamp-on ammeter, fan-guard probe, surface roughness gauge, feeler gauges, and the high-potential tester.

INTERNAL THERMAL overload sensing devices (right) are installed in all explosionproof motors. Goodwin Pray plant manager shows how devices are installed—one in each phase, in series with control circuit leads.

200-HP MOTOR drives huge sand-processing fan via belt/pulley drive at left. The Davis troubleshooting team is using portable vibration analyzer/balancer to pinpoint cause of excessive vibration that could lead to ultimate breakdown of equipment.

Portable instruments speed electrical troubleshooting

Two expert servicemen reveal how they use portable instruments to speed in-plant electrical troubleshooting.

FAST, efficient troubleshooting is vital to the continued success of H. M. Davis & Sons, Inc., an apparatus service firm of Bridgeton, N. J. The company specializes in the repair and rebuilding of electrical and mechanical equipment; however, troubleshooting and maintenance of in-plant equipment is an essential service provided to many of their customers.

The firm's top field men are Bob Davis and Wayne Lang, who point out that proper use of portable test instruments is the key to effective troubleshooting. After years of experience they have found the most useful instruments to be the multitester, the snap-around volt-ohm-ammeter, and the vibration analyzer/balancer. The following typical trouble calls show how they use these instruments.

Intermittent shutdown of exhaust fan motor

At a large sand-processing plant, a 200-hp, 460-volt induction motor driving a huge exhaust fan had been kicking out intermittently, interrupting the continuous sand-processing operation. The resulting loss in production was costly, so plant management called in the Davis firm to solve the problem.

When Davis and Lang arrived at the plant, they first made a visual inspection of the motor, which was driving a large 6-ft-dia fan via a multiple belt drive. With the motor running, they checked the assembly for signs of malfunction, such as unusual noises, hot bearings, loose mounting bolts, slack belts, or any sign of accidental damage to enclosures.

A similar inspection was made at the autotransformer-type reduced-voltage motor controller. With the motor-circuit disconnect switch off and locked out, they checked for loose main fuseholders, for any darkened or burned spots, appearance of overload relays, condition of "start" and "run" contactors, and physical appearance of both line and control conductors.

According to Davis, these inspections, which take only a few minutes, often indicate the source of the trouble. However, in this instance all components checked out OK. Therefore, the next procedure is a check of circuit conditions using portable test instruments. First, the motor was started and the running current was measured using a snap-around ammeter. Davis emphasizes that common sense is important to obtain consistently accurate readings with the snap-around ammeter. For example, be sure that stray magnetic fields do not affect the current reading. Arrange other conductors so that they are as far removed as possible from the conductor under test. If testing is being done at a control panel, try to take the current readings at a location remote from the relay magnet coils. For the same reason, avoid taking current readings on conductors at a point close to a transformer. All these influences can affect the accuracy of the reading.

The motor running current in this case, as measured by the snap-around ammeter, was 265 amps. This was significantly higher than normal (225 amps) and indicated an overload. This level of current could cause slow heat-

RUNNING CURRENT of 200-hp motor is easily checked with snap-around ammeter. Wayne Lang, service technician for H. M. Davis & Sons, Inc., records readings taken on all three phases for future reference. Motor controller is a size 5, 480-volt, autotransformer reduced-voltage type.

ing of the overload elements, and tripping could occur during a line voltage, switching or machine-starting surge.

Before beginning the search for the overload, Lang suggested that the supply voltage be checked, since leads could easily be inserted into the ammeter to permit it to be used as a voltmeter. This was quickly done and they proceeded, secure in the knowledge that supply voltage was normal (460 volts, balanced on all three phases).

Davis and Lang explained that an overload (indicated by the higher-than-normal current) is usually caused by a change in the driven load or a mechanical malfunction in the motor itself.

Because the load in this instance was a fan, it was unlikely that the load would be responsible for additional torque. On the other hand, the many components in the drive system suggested several possible sources of trouble. Lang explained that a bad bearing, bent shaft, misalignment of sheaves, or imbalance on the fan blades, shaft, or sheaves are common causes of overload on large complex machinery.

To quickly pinpoint this kind of problem, the Davis team uses a portable vibration analyzer/balancer. This compact instrument identifies vibrations and displays their amplitudes and frequencies on meters when a cord-connected sensing head is in contact with the rotating apparatus. The unit may be battery-operated or plugged into a 115-volt source.

A series of curves furnished with the tester, on which vibration amplitude is plotted against the frequency of vibrations, aid in diagnosing the conditions encountered. However, Davis is guided more by previous readings taken after the unit was aligned and balanced properly. Based on experience with similar equipment, Davis points out that vibration amplitude should not exceed 2 mils at any of the measurement locations. (Other types of equipment may have limits as low as 0.1 mil, depending upon speed of rotation.)

The frequency of the vibration helps to pinpoint the source. For example, a frequency equal to the rpm of the motor indicates that trouble could be caused by eccentric journals, misalignment, or a bent shaft. A frequency twice the motor rpm indicates

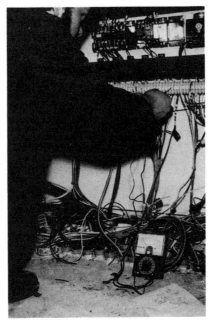

MULTITESTER, set to operate as voltmeter, reads zero voltage at terminal point (test point F), indicating location of problem. Closer examination revealed wire had burned and broken loose from terminal.

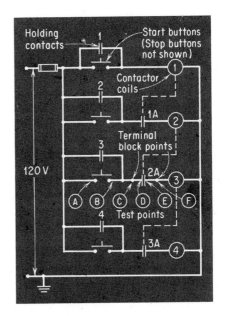

SIMPLIFIED DIAGRAM shows key control elements in sequential motor-starting scheme. Voltage checks were made at points A through F with multitester to solve control-circuit malfunction.

misalignment, rubbing, reciprocating forces, etc.

The vibration analyzer indicated an excessive vibration (4 mils) and that an imbalance somewhere in the fan blades was causing the problem. Using a vector analysis technique, Davis and Lang were able to pinpoint the exact location of imbalance on the 6-ft-dia multiple-blade fan and add weights in the proper location to provide a precision dynamic balance.

After this was done, proper readings were obtained on the analyzer (less than 2 mils for a fan rotating at 750 rpm), and a check at the motor starter with the snap-on ammeter showed that motor running current was normal.

Troubleshooting control circuits

Davis and Lang agree that their multitester, sometimes called a multimeter or volt-ohm-ammeter (VOM), is an especially useful instrument. This highly versatile tester is capable of measuring voltage, resistance, and low values of dc current. Accuracy of the meter is dependent on the amount of current required to drive the pointer or needle full scale—the smaller the current, the more accurate and sensitive the meter. Generally, sensitivity (which determines accuracy) is expressed in ohms per volt.

This represents the impedance the meter presents to the circuit when used as a voltmeter. Multimeters are rated 1000, 5000 or 10,000 ohms per volt ac. Meters having a sensitivity of 10,000 ohms per volt have higher accuracy because their higher impedance creates less load on the circuit under test.

Davis emphasizes that safety must always be considered in the use of the multimeter. Know the voltage levels and shock hazards related to all equipment to be tested. Never try to take voltage readings on power distribution circuits rated over 600 volts. Measurement of high voltages is accomplished by installed instrument transformers and meters.

When making voltage measurements on power and control circuits, be sure the meter selector and range switches are in the correct position for the circuit under test before applying test leads to the circuit conductors. To prevent damage to the pointer, always use a range that insures less than full-scale deflection of the pointer. A mid- to ⅔-scale deflection of the pointer assures the most accurate readings.

When operating the instrument as an ohmmeter, they find that it is particularly useful for checking de-energized control circuits and taking resistance readings of circuit compo-

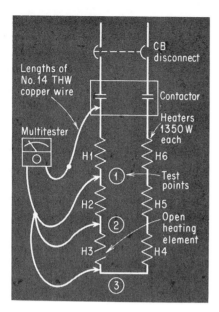

PROCESS HEATER CIRCUIT consists of six 1350-watt strip heaters controlled by contactor and protected by 2-pole circuit breaker. Open heating element was found by using multitester set to operate as ohmmeter and making checks at points 1, 2 and 3.

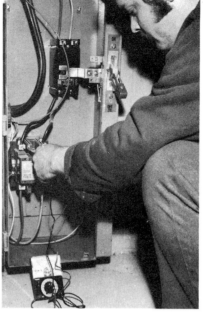

VOLTAGE AT CONTROLLER to strip heater circuit is checked with multitester. Tester was then used as an ohmmeter to check continuity and measure resistance of heater circuit.

causes high resistance, which results in heat, expansion and contraction, more loosening, carbon buildup, and finally higher resistance and excessive heat that causes complete destruction of the termination.

Checking circuits of process heaters

The multitester, used as an ohmmeter, is particularly useful in determining the condition of process heater circuits. In one typical application, strip heaters were installed in an enclosed conveyor to heat materials as they were carried between buildings. Six heaters, with a total rating of 8100 watts, were connected in series and powered at 480 volts. After a period of time, it was suspected that the heaters were not producing their full rated heat. Using the multitester, Davis first checked the supply voltage to the heaters at the control contactor. The voltage supply measured 465 volts, which was considered adequate for proper operation. Then he decided to make a continuity check to see if the circuit was open or shorted at some point. A quick calculation indicated that proper resistance of the circuit should be about 28 ohms plus a few ohms for conductor resistance, so Davis set the multitester to operate as an ohmmeter. A check at the contactor showed an infinity reading, indicating that the circuit was open at some point.

According to Davis, the fastest way to locate the open was to check the circuit continuity from its origin (controller) starting at one leg and making checks at each heater terminal in the series circuit. Because the controller was located nearly 50 ft from the heater installation, he had to improvise to create leads long enough to reach the points to be checked. He did this by attaching a 70-ft length of No. 14 THW copper wire to each multitester lead. One lead was connected to one circuit leg at the controller (see diagram); the other was used as a probe lead. As shown in the diagram, he took measurements at point 1, where he obtained a proper reading of about 5 ohms; at test point 2, he obtained a reading of approximately 10 ohms. At test point 3, the multitester showed a reading of infinity, indicating that the open was between test points 2 and 3. Suspecting that heater H3 was at fault, he checked across the terminals of the heater with the mul-

nents. These readings can be used later for comparison during troubleshooting.

The multitester proved its worth recently when Lang was called to solve a control circuit problem. The control system required that four motors start in sequence—that is, when start button No. 1 was pressed, motor No. 1 would start; when start button No. 2 was pressed, motor No. 2 would start; etc. The scheme did not permit No. 3 motor to be started before No. 2, nor motor No. 2 before No. 1, etc.

In this instance, the operator was able to start motors No. 1 and 2 but when he pressed button No. 3, motor No. 3 would not start. When Lang arrived at the plant, he made a visual examination of motor No. 3 and its starter. Both appeared normal. Then he obtained a schematic of the control system (see accompanying simplified diagram). He noted that when pushbutton No. 1 was pressed, coil No. 1 would energize, holding contact (1) closed to hold the coil energized, and line contacts to the motor would close, energizing the motor. In addition, coil No. 1 was furnished with an auxiliary contact (1A), which was wired in the start circuit to coil No. 2. This meant that the contact 1A had to be closed in order for current to flow through to coil No. 2. A similar arrangement existed for the other circuits.

Now that he understood the operation of the control system, he naturally directed his attention to the circuit supplying coil No. 3. He decided to make a voltage check of the circuit using his multitester (see diagram). First, motors No. 1 and 2 were started. Then, Lang checked to see if voltage was reaching the line side of start button No. 3 (test point A). Proper voltage was present here. With the operator holding the start button down, he checked for voltage at load-side terminal of the start button (test point B). This checked out OK. The circuit from the start button to contact 2A was brought to a terminal block. A voltage check here (test point C) proved to be OK. Voltage also was present at both sides of the auxiliary contact 2A (test points D & E). From contact 2A to the coil No. 3, the circuit was routed through a terminal block. At this location (test point F), he could not obtain a voltage reading. Closer examination of the terminal point revealed heavy oxidation and burned copper. The circuit was open at this point; thus, coil No. 3 could not be energized to start motor No. 3. A new termination was made up and the system operated normally.

Lang explained that the probable cause was a loose connection which had existed from the time of original installation. A loose connection titester still set to operate as an ohmmeter and he obtained an infinity reading (open heater).

Next, he removed the faulty heater, installed a new one, and made a continuity check at the controller. The proper reading was obtained (approximately 28 ohms). To be certain that the circuit had not been grounded accidentally during testing, he checked resistance to ground with the multitester set on its highest ohmmeter scale (R × 10). The reading was infinity, as it should be. Finally, he energized the circuit and took voltage readings. A copy of the recorded voltage, current and resistance readings was sent to the plant maintenance engineer for reference during future troubleshooting or preventive maintenance procedures.

Troubleshooting motor capacitors

 A COMMON problem encountered in the service end of the electrical construction industry is troubleshooting capacitor-type motors. What is a fast, sure method of checking the motor capacitor?

 **THERE ARE** a number of procedures for testing capacitors used with single-phase capacitor-type motors. Here are a few.

Method 1. One way to solve this problem is to replace the capacitor with a known, good capacitor. If the motor runs, the original capacitor was at fault; if not, the trouble could be the capacitor or elsewhere in the motor.

Method 2. Appliance servicemen check capacitors quickly with an instrument known in their jargon as a "cappy." This instrument is actually a bridge-circuit tester that indicates whether or not the capacitor is good.

Method 3. Another fast method of testing a capacitor is with a multitester using the ohms scale. With the capacitor out of the circuit, place the ohmmeter leads on each terminal of the capacitor. If no reading is obtained, the capacitor is open. If an immediate reading of continuity is obtained, the capacitor is shorted. If the pointer jumps up to approximately mid-scale and slowly returns towards infinity, the capacitor is probably good. (Since the ohmmeter impresses only 1.5 volts dc on the capacitor, it may check out good on this test. However, it may break down when placed on line voltage. Line-voltage tests are described in the following two methods.) Also check whether or not the capacitor is grounded by placing the ohmmeter leads between one terminal and the capacitor case. A zero ohms reading indicates a ground.

Method 4. Connect the capacitor in series with a 10-amp fuse across a 120-volt, 60-cycle ac line as shown in sketch A. If the fuse blows, the capacitor is shorted and must be replaced with a new one. If the fuse does not blow, the capacitor will be charged up to some voltage. This charging will only take a few seconds. If the line is then disconnected from the capacitor and the capacitor terminals are carefully shorted by a screwdriver, a spark will be drawn at the terminals (sketch B). If a capacitor has wire leads, the leads can be touched together. If a spark cannot be drawn, the capacitor is either open or has decreased in capacity. Test should be repeated a few times.

Method 5. A reliable method of testing a capacitor is shown by the circuit in sketch C. The fuse should be sized slightly higher than the capacitor-rated current. The object of this test is to obtain the capacitor's microfarad value by measuring voltage and current in the circuit and inserting these values in an established working formula. The capacitor should be energized only for a short time because some capacitors are rated for intermittent duty. Make measurements quickly using a snap-on volt-ohm-ammeter to obtain voltage and current readings. Then the approximate capacity in microfarads can be computed by substituting the readings in the following formula (a 60-Hz supply is assumed):

$$\text{Microfarads} = 2650 \times \text{amps} \div \text{volts}$$

This computed value should be compared with the value marked on the capacitor. If the capacitance in microfarads is not within 5%, the capacitor should be replaced.

SKETCH A shows circuit used to test capacitor for a short. In sketch B, screwdriver shorting leads of charged capacitor should draw a spark.

Safety rules should be observed at all times during tests. Sketch C shows test circuit used to determine capacity of a capacitor.

Checking capacitor motors

A systematic analysis of the circuits and components of these popular motors, together with a few simple tests, can greatly simplify troubleshooting.

By J. L. WATTS, A.M.I.E.E.
Southampton, England

THE FAILURE of a capacitor motor to start or function properly does not necessarily indicate a defective winding. Other components could be at fault. These include defective starting relays or switches or their mechanisms, faulty capacitors or autotransformers, loose connections at built-in thermal overload devices, and bad bearings on the motor or the load it drives.

Faulty motors should be checked systematically to pinpoint the trouble quickly. In general, there are two parallel circuits through a capacitor motor at standstill, so there should be a noticeable hum when the motor is switched on. If no hum exists, the first obvious check is to be certain that the branch-circuit protective devices or control switches are not open. The next step is to connect a test lamp across the line terminals of the motor (P1 and T4 in Fig. 1). If the lamp fails to light, there is an open circuit in the supply to the motor. The motor should function properly once the open has been located and corrected. However, if the lamp lights (on terminals P1 and T4) but the motor doesn't hum, the cause is either a break in the line conductors *inside* the motor or an open circuit in *both* the starting and running circuits. The most likely point of such an internal open circuit is a thermal protector with open contacts due to overheating or faulty starting. Some protectors may reset automatically after cooling down; others may have to be reset manually. Another cause of internal open circuits could be faulty thermal protector contacts or connections indicated by failure of a test lamp to light when connected across terminals T1 and T4.

Improper starting under load

If the motor hums but refuses to start or start properly when energized, the supply voltage may be low, or a dual-voltage motor might be improperly connected with its windings in series on the lower voltage of its range. Obviously, wrong hookups would be suspected only at the initial startup. If the line voltage and connections are in order, the motor should be uncoupled from the drive and energized at no load. If it starts correctly, faulty starting on load may be due to (1) overload, (2) a ground fault or short circuit in the windings, (3) a defective capacitor or autotransformer, or (4) an open circuit in the starting capacitor of a two-value capacitor motor. A motor also could start properly unloaded but fail to start under load if the bearings are badly worn and cause the rotor to rub on the stator core.

If the rotor clears the stator core and the motor is self-starting when operated unloaded, its performance should be observed. If it shows signs of overheating in a short time, it should be turned off. Contacts of a starting relay or centrifugal switch sticking in the closed position could cause the overheating. If so, the overheating should cease after one lead of the starting winding is disconnected after the motor is started and up to speed. Replacing the switching mechanism should correct the problem.

Overheating of an autotransformer used with some two-value capacitor motors may be due to a short circuit in the capacitor or a short circuit or ground fault in the autotransformer. One of the latter two would be indicated if the overheating continued after disconnecting the capacitor.

Testing windings

If the motor overheats when running at no load, after one lead of the

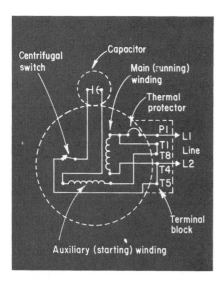

FIG. 1. Connections are shown for a typical single-voltage, capacitor-start motor. Note that there are two parallel circuits while motor is at standstill. Terminal numbering shown is for motors marked according to ASA designations.

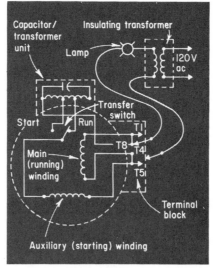

FIG. 2. Testing for a ground fault in a two-value capacitor motor is done simply by a line-voltage lamp test. Placing probe X on terminal T8 and probe Y on metal frame of motor will detect presence of ground in running or starting circuits. If lamp lights, removing terminal jumpers T1-T8 and T4-T5 in turn will indicate which circuit is grounded.

starting winding has been disconnected, a ground fault or short circuit in the windings is indicated. To test for a ground fault, a line-voltage test lamp and a 1:1 double-wound transformer may be used as in Fig. 2. To test for a short circuit between main and auxiliary stator windings, the two windings may be disconnected from each other and the test probes X and Y applied to one end of each of the two windings. To test for a short circuit in a main winding, the dc resistance of similar sections of the windings can be measured and compared—a relatively simple matter with a dual-voltage motor. Unequal resistances in similar sections indicate a short circuit.

A two-value capacitor motor with an autotransformer may fail to start loaded and yet may start unloaded if the starting capacitor is open-circuited. In this case, it is probable that starting will be improved if the capacitor is bridged by a lamp or shorted by a piece of wire. Other capacitor motors may start properly only on no load if the starting capacitor is shunted. For these, starting may be improved if the capacitor is disconnected and the external leads of the capacitor are bridged by a suitable lamp. A short circuit in an electrolytic capacitor may be indicated by the top of the capacitor having been blown off by a rise in internal pressure.

Checking capacitors

A capacitor may be tested for a short circuit by connecting it across the ac supply for a few seconds through a fuse rated slightly over the normal capacitor current. The fuse will open if the capacitor is short-circuited. On a 110-volt, 60-cycle supply, the normal current of a capacitor may be considered as 0.4 amp per 10 microfarads, or 0.8 amp per 10 microfarads on a 220-volt, 60-cycle supply.

However, a better test for an open circuit, short circuit, or loss of capacitance (as can occur if the electrolyte in an electrolytic capacitor dries out) is shown in Fig. 3. In this test, the capacitor should be energized just long enough to take quick voltage and ampere readings, because a capacitor of this type is designed only for intermittent duty. On a 60-cycle ac supply, the capacitance in microfarads may be taken as equal to $(2650 \times amps) \div volts$. This calculation should be com-

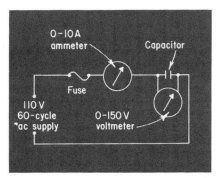

FIG. 3. A simple test to determine the capacity (in microfarads) of a capacitor is to insert an ammeter and voltmeter in the circuit as shown. The capacity rating on the capacitor should be approximately equal to $(2650 \times amps) \div volts$. If not, the capacitor should be replaced. Shorted, open, or low-capacity capacitors are common defects found in capacitor motors.

pared to the value marked on the capacitor. Any replacement capacitor should be approximately the same rating as the original one to insure the required starting torque.

Failure to start at no load

If a motor hums but fails to start when switched on unloaded, the starting capacitor should be bridged by a lamp or, momentarily, by a piece of wire. If this starts the motor, there is an open circuit in the capacitor. If it does *not* start the motor, the rotor should be spinned rapidly by hand and the motor then turned on. It probably will accelerate in the direction it has been turned. If it runs up to full speed in a normal manner, an open circuit in the starting circuit should be suspected. But if the motor reaches only about ¾ normal speed, the running winding has an open circuit. Fig. 4 shows a method of using the test set in checking for an open starting circuit. The same method can be used to check for an open circuit in the main winding. Note that the two windings must be disconnected from each other for these tests, and the capacitor must be bridged when testing the starting winding circuit.

An open in the starting winding circuit often is due to the contacts of a centrifugal switch or starting relay not closing properly at starting. This would be confirmed if the motor started properly only when these contacts were bridged by a piece of wire. Where there is excessive end play in the rotor, the motor may stop at

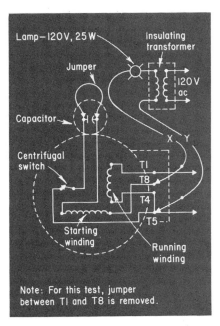

FIG. 4. Testing for an open in the starting circuit of a capacitor motor involves placing probe X on terminal T8 and probe Y on terminal T5. Notice that the jumper between T1 and T8 has been removed for this test and that the capacitor leads are jumpered. If there is no open circuit, the lamp should light.

times if the centrifugal device moves away from the contacts mounted inside the motor case. Such a condition may not permit the contacts to close once the motor returns to standstill. In this event, the trouble may be overcome by fitting washers onto the shaft in order to take up excessive end play.

An electromagnetic starting relay must be adjusted to close its contacts when the motor is turned on at rest with the minimum supply voltage likely to be experienced, and to open its contacts as the motor approaches its working speed—even when driving its rated load with maximum and minimum values of anticipated supply voltage.

If a motor overheats on load, it may be due to overload, low voltage, one or more windings shorted or grounded, an open-circuited winding in a dual-voltage motor operating on the lower voltage, incorrect connections of a dual-voltage motor, a short-circuited capacitor of a permanent-capacitor motor or a two-value capacitor motor, or faulty contacts of a centrifugal switch or starting relay. A systematic approach to the problem will uncover the defect as indicated by the previously described tests.

Identification diagrams speed troubleshooting

UNSCHEDULED DOWNTIME is a major concern in most industrial plants, especially in continuous-process facilities where an interruption in production can cost thousands of dollars. Generally, these facilities employ skillful electrical people who know the latest troubleshooting techniques and have access to modern tools and instruments to assure efficiency.

Wiring diagrams, schematics and related data are essential to the troubleshooting process. However, for maximum effectiveness, a total plan identifying every component in complex interrelated power and control systems should be followed. This plan should consist of a set of documents—completely correlated drawings, schedules and procedures to simplify equipment and system operation and speed circuit tracing. To accomplish this, the following are required:

1. One-line diagrams, schematics, internal wiring diagrams and external connection diagrams for each system.

2. A tabulated cable schedule showing the identification number, routing, number of conductors, and rating of each cable in the plant.

3. A complete set of conduit drawings, with the cable numbers marked on all conduit runs.

4. An identification label for each piece of equipment noting its relation to the process.

5. Clear and permanent identification of every wire at every termination in every circuit.

6. Careful supervision and inspection during equipment installation or alterations to be certain that identification labels as required are properly installed. It is highly recommended that the identifications on the equipment and those shown on working drawings, shop drawings, and troubleshooting drawings be thoroughly checked to be certain that all are in agreement.

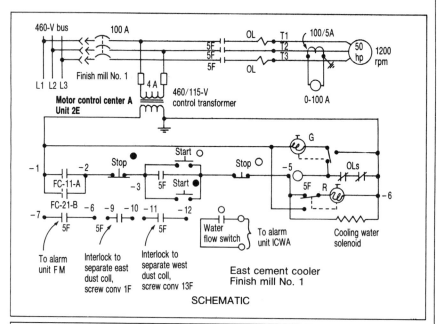

SCHEMATIC

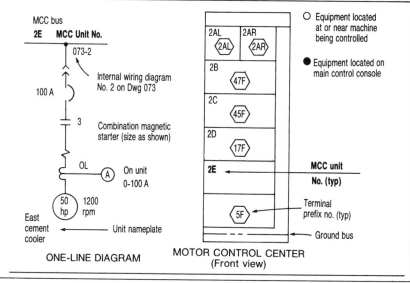

ONE-LINE DIAGRAM

MOTOR CONTROL CENTER (Front view)

DIAGRAMS incorporate identification technique and show circuits pertinent to a 50-hp, 440-V motor on a cement cooler in a finish mill. On the schematic, note that the motor starter is identified as unit "2E" in motor control center "A". The "2E" identification is also noted on the one-line diagram and shown on the front view of the motor control center. This indicates that the unit is in space E on the 2nd section of the motor control center. For continuity, the same "2E" identification appears on all other diagrams. Note use of symbols (open or closed dots) to indicate location of components. This helps the troubleshooter to find these components quickly. On the motor control center diagram, the hexigon containing the number "5F" identifies all terminal blocks, connections and equipment relating to that particular motor. This identification is carried through on all drawings and installed equipment. For example, the interconnection cable between the motor control center and the main console will have control wires identified with a prefix of "5F."

INDEX